natural :

*the african wanderings
of a bemused naturalist*

natural selections

*the african wanderings
of a bemused naturalist*

DON PINNOCK

DOUBLE
STOREY
a juta company

In association with
GETAWAY magazine

First published 2002 by Double Storey Books,
a Juta company, Mercury Crescent, Wetton, Cape Town

© 2002 Don Pinnock

ISBN 1-919930-03-5

All rights reserved

Permission to reproduce any of these essays in a magazine rests
with *Getaway*, to whom application should be made

Cover design: Julia Raynham
Text design and layout: Sarah-Anne Raynham
Printing by: ABC Press

contents

PREFACE   VIII

1. WILD TRAVEL   2
    Notes from heaven   *4*
    In search of the strange   *10*
    Ballets of catastrophe   *18*
    Paddling back in time   *23*
    Travelling the trails of !Xam   *30*
    Above the sea of mountains   *40*
    Into the sands of silence   *47*
    Naturally Nyalaland   *56*
    Absolute Orange   *61*
    Into the forbidden desert   *81*

2. FLYING THINGS   92
    Superpilots   *94*
    Our one big mistake   *100*
    Parliament's poached penguin eggs and the Great Guano War   *108*
    Angels of the night   *115*
    Balancing acts on the edge of extinction   *123*
    Sinking gannets and the mystery of a feather   *130*

3. SWIMMING THINGS 136
   Everything but an eye-biter *138*
   Lords of the sea *144*
   Snail hunting in the mother lake *156*
   Neptune's little ponies *164*
   Monster hunting in the Okavango *169*
   Wild orchestras of the night *176*

4. WALKING THINGS 182
   Ships of the desert *184*
   Wild steeds of the desert *190*
   Monkey business *196*
   Lords of the trees *205*
   The long, dark road of the cave cricket *212*

5. HOMO SAPIENS OF DISTINCTION 218
   Worker in a science yet unborn *220*
   Caves of antiquity *227*
   The madness of a collector *234*
   I look your mouth and it be sweet *241*
   Some lessons on dreaming the world *250*
   The river makers *256*

6. STRANGE STUFF 262
   Galactic ecology *264*
   The strange history of a fix *271*
   God dam crazy *279*
   Just sniffing around *288*
   An enthusiasm for fire *294*
   Of rain and reflections *301*

preface

It began, we're told, from nothing. From a single point in the vastness of … what? Everything exploded into existence. Astonishing. An unimaginable woosh of energy that ended up as galaxies, stars and us. And elephants and macaroons and lemurs and …

While flying across winter clouds roiling up over the Drakensberg and eating a soggy Sky Chefs cheese roll one day, it hit me: we are so fortunate. We're the only creature we know of that can look at this unfolding universe and the forces that hold it together with a measure of understanding – and appreciate its head-spinning beauty. How privileged we are, and what a terrifying responsibility. I shivered and pulled down the blind. But the size of the woosh wouldn't go away.

My problem with our place in all this vastness goes back a bit. Around the age of 12 I read Robert Ardrey's *African Genesis*. It blew me away. Whereas my friends dreamed of flying jets, farming sheep or shooting hapless elephants in Kenya, I fantasised about discovering the Missing Link alongside Louis Leakey in Olduvai Gorge, or finding a full skeleton of *Doedicurus* in the koppie above my house in the Eastern Cape.

What fascinated me was palaeontology's vast timescale. If it took 15 billion years to fashion the universe and a million to make *Homo sapiens*, then every bug or butterfly was an absolute treasure. (It was a perception of just how long life took to create the creatures around us that brought Alfred Wallace and Charles Darwin to an understanding of natural selection.)

As it happened, I took off on a different trajectory, becoming an engineer, then a journalist, a criminologist, an historian and a specialist in aberrant teenage behaviour – in that order. I wrote a clutch of theses and books, most of which now give me the sense of having been written in some previous, unimaginable life.

Then – serendipitously – I was invited to become a travel writer and photographer for *Getaway* magazine. My beat: all of Africa. I couldn't believe my luck.

Roaming the continent, camera bag over my shoulder and notebook in my pocket, it soon became clear that Africa was far more than political skulduggery, starving refugees and mind-numbingly overcrowded cities. It was also full of the wackiest creatures in some of the most wonderful places on earth.

Along the way I re-read Ardrey's *African Genesis*, discovered, a fascinating book on island biogeography – *The Song of the Dodo* – by David Quammen, and nosed my way into the controversial natural history essays of Stephen Jay Gould. Marvellous stuff. It was while hiking up through the ancient rainforests of Amber Mountain in northern Madagascar, that it occurred to me I'd had fallen hopelessly in love with Planet Earth (imagine if we'd got Mars or Pluto?).

David Bristow, the editor of *Getaway* and a perceptive fellow indeed, noticed my growing enthusiasm and suggested I write a monthly column on natural history. This was appealing but daunting: half a million well-educated, highly literate people read the magazine and I had no scientific background. Then I discovered that Ardrey didn't have one either – he was a playwright – and Quammen has a literature degree from Oxford. So background didn't necessarily disqualify one from writing popular science.

That's how the column, *Natural Selections*, was born. I was to find that having the wrong background was something of an advantage. Scientists tend to hunker down in their disciplines and write only facts as they see them. Dilettantes like me could speculate, and move more freely between disciplines without feeling like a fraud.

So this book is a collection from the column as well as from several other sources. Its arrival in print is thanks to the many naturalists, scientists and wilderness guides who led me to wonderful places, fielded dumb questions with good grace and taught me about nature's many quirks. They dredged some startlingly large questions from things ordinary folk pass over with hardly a glance.

I am also profoundly grateful to that walking encyclopaedia, *Getaway* editor David Bristow, friend and mentor, who pressed the 'Go' button on the column and ensured that it all made good sense. And to my wife and favourite author, Patricia, who checked every word with a red pen in her hand, and to my children, Gaelen and Romaney-Rose, who thought having an oddball adventurer for a dad was cool. Thanks, guys.

wild travel

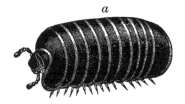

a

Notes from heaven

The ripple of frogs counterpoints a night so still the ants seem to be walking on tiptoes. High overhead, tamboti and knobthorn trees are catching stars and a comet or two in their interlaced branches. And below them three logs consult head-on to produce a flame which urges the kettle to greater effort.

It feels good to be down on the naked skin of Africa in the small hours. I'm on night watch, probing the perimeter with a torch somehow less bright than my imagination, peering for predators and unwelcome ungulates: wishing them absent; hoping they're there ...

At the outer edge of this holy night a lion offers a purring growl and a hyena responds with a lilting whoop – shock reminders to remain vigilant in a place where we are both intruder and prey.

The skein of romance suddenly unravels and it occurs to me: this is crazy. My fellow travellers are fast asleep and I'm alone on night watch in the middle of a Big Five wilderness area with nothing but a small fire and a torch. In the ear-zinging silence the noise of my jacket sleeve brushing against my side makes me jump – I can hear leaves dropping from the trees.

Circling the crude camp beneath the combretum canopy, I notice worn bark where rhino, buffalo and elephant have scraped dried mud and ticks off their hides. It gives me the eerie feeling of trespassing on a ritual site.

The Big Five were named because they are the most dangerous animals to hunt – and they are all here.

A deep out-breath makes me jump, torch probing the moonless dark. A glint of reflection, then two eyes too far apart to believe, are swaying rhythmically like red fireflies and coming my way.

My heart takes on a rhythm of its own: it's a huge buffalo heading straight into camp. Buffalo are smart and fear little. Old *daga*-boys cast out of the herd can be mean and this one doesn't look as though much will deflect its chosen path. It's a case for the professionals. I shake wilderness guide Mdicini Gumede awake.

He leaps up out of sleep with an energy surprising for a man in his 70s. For a heart-stopping moment the old Zulu and the old bull walk towards each other. Waving his arms in the air Gumede shouts: '*Sawubona Nyati*' (Greetings, Buffalo).

The great beast stops in its tracks and swings side-on. In my torch beam it looks as big as a pantechnicon.

'*Hamba mfwetu, hamba*' (Go, my brother, go).

The buffalo concedes with a deep out-breath and ghosts away as silently as it came. Gumede grunts approvingly and goes back to bed, leaving the night to me and a far-off, whiffling hyena.

I stack up the fire and sit reflecting how much better it is to greet a buffalo as a brother than to line it up in the sights of a gun.

When dawn comes it's initially the domain of the birds. The francolin is first to crack open the rim of the sky, letting in a soft glow which backlights a slow-dancing marula tree doing t'ai chi.

A brace of hadedah opens the crack further, letting in red rays under the grey cloud blanket, then a brown-hooded kingfisher takes up the task, peeping urgently at the lightening sky. This wakes up a bulbul and tips off a ripple of tweets and pips all round. A pair of giant hornbills begin a duet: '*Ngududu,*' starts the male, and the female answers: '*dudududu.*'

Wings begin to unfold, criss-crossing the woodland with *prrps* of feather-flight. The racket disturbs a grumpy hyena which *whoop-hoops* in protest, then settles down to an uneasy slumber after a night on the prowl.

The next shift belongs to the baboons.

'*Baa-hoo, ba-hoo?*' the alpha baboon along the cliff opens his day's debate.

'*Oo-ha,*' begins a female, but a cheeky youngster interrupts: '*chaar chaar chaar.*'

'*Ba-hoo?*' asks the male again, sharply. But before his mate can answer, the youngster is at it again: '*Chaar-chik, chaar chaar.*'

The male considers this for a while, then decides he's not going to be upstaged by a junior: '*Baa!*' he yells. '*Baa.*' Thump, smack. '*Baa.*'

'*Waa, waa, waaaaa,*' shouts the upstart and whimpers into silence.

'*Ba-hoo?*' the male tries again.

'*Oo-ha,*' comes the soft answer. But it's too late for a quiet adult conversation: the whole troop is awake.

'*Scree. Chirr chirr. Ba. Gwaa. Chee ...*'

For the baboons and the rest of the river-bend community another day has begun. It's a signal to roll out of our warm bags and think about breakfast. The day before, wood – fallen, not broken – had been gathered on trail: knobwood, wild camphor, bush willow and rhus for cooking, tamboti for the long night watch.

Sand had been piled in the centre of the camp, then flattened – protecting the delicate roots undergirding the clearing. Leaves had fed the first spark, then twigs, then the head-end of three logs had been poked in to hold and grow the flame. On top had come the trusty tripod: a neat, simple brace for kettle and pot.

In the morning the flames are coaxed to new heights and we gather sleepily round the kettle needing warming tea. Afterwards the embers are allowed to die. Water thuds out the last heat and the grey amalgam of wood and brown sand is mixed like *dagha*, tossed into the cooking pots and carted away to be chummed back to earth.

Pieces of wood which remain are picked out of the sand and removed. Afterwards a twig with leaves is used to sweep the clearing, leaving nothing but footprints.

Memories of the magic night are soon obliterated by searing light. Boots follow boots down the Mfolozi River. In a daze of heat in black rhino country we become one creature, sticking close together – eight packs, 16 boots and one mind. When the back stops the front knows and stops also. The Zulu say that midday – when animals stand in the shade and dream – is a time when their spirits are abroad.

Gumede's bootfalls ahead of me are hypnotic: heel-toe-up, heel-toe-up. At a rhino midden, datura is growing – *malpitte*, used to invoke the water spirits.

The sharp warning bark of a big chacma baboon breaks the spell. The line suddenly falls apart: eight people again, some looking at the hilltop, others at the midden. In the riverbed two giant buffaloes gaze up, wondering, no doubt, what all the fuss is about. We emulate their oxpeckers in the ritual of lunch.

Before us lies the river, lazy and unbridged: the only way over is through. There's nothing like wading waist-deep through a crocodile-infested river to sharpen your attention and tone up your peripheral vision. Crocs don't show up easily in muddy water but they've got specialised equipment to see you and smell you without being detected. The scene is deceptively calm – a wide bend in the river and sand banks without worrying slide-marks.

'They can take buffalo,' says wilderness guide Ian Read helpfully. 'Big ones here.'

We make it, of course. But for the next hour I can't get the thought of long rows of teeth out of my mind. Somehow it seems better to be eaten by a mammal.

On a silver bend of the river we find rhinos at last. Two are white and two are black – but they all look like the 'before' part of a Surf advertisement: muddy brown and smeared. One has a horn so long she could unseat a jousting knight, but instead she browses and defecates with stolid practicality, unaware of our presence.

An old buffalo notices us, though, and walks forward, his wet cow-nose in the air and sporting a massive boss which looks well battered. He takes his time, unhurried by us monkey creatures on the cliff edge.

As the golden sun boils gently downward through wraiths of cloud at the edge of the day it occurs to me that, compared with the bovine creatures of the wilderness, we move like a video on fast-forward.

The rhino munch-mow in methodical loops; the buffalo goes about its business with cud-chewing contemplativeness; giraffe select tree-tops with studied delicacy.

But to the parkland trees even these graceful creatures must seem no more than a blur. And least speedy of all is the buffalo thorn *(Ziziphus mucronata)*, for the ziziphus is a slow dancer.

If you were to watch it for a hundred years or so you'd see it thrust to the left with an energetic limb. The whole tree would lean to balance the lopsided load, then recover by remembering an alternative solution: another limb would unfold in the opposite direction, this one a ground-level right which would curl upward as the left curled inward.

Given, perhaps, a few decades you'd witness a third limb exploring upward from the base, arching forward then swooping left and right again, perfectly balancing the incongruent flight of the first two.

As an afterthought of this many-limbed jazz-dance, leaves would sprout from all the upper extremities, uniting the whole creature into a tree of powerful presence.

The Zulu consider it a kindred spirit, calling the tree *Lahla inkosi* (fetch the king). They believe that if a loved one dies far away and the body cannot be returned home, the spirit will leap into a ziziphus branch laid beside it. It is often told that members of the family coming home with the spirit-filled bough will buy two tickets on the train – one for themselves and one for the ziziphus.

As night falls again over the Umfolozi wilderness area and a fire crackles into life we busy ourselves with food and friendship. Some prepare for sleep, others scan the perimeter for danger. We humans have a far longer history of life lived this way than in buildings and cities. It feels good to be back at ground level, it's a place the soul understands intuitively. As the day fades to black, a lion greets the stars with a roar that shakes the kettle lid.

In search of the strange

'The way to cook a tenrec', Zakamisy explained as we stomped up through the dripping mist forest, 'is first to boil it, then peel it to remove the prickles. After that you grill it.'

Tastes a bit like pork, evidently.

The conversation about tenrecs had begun back in Antsiranana, where Zak said he thought he could root one out from the underbrush. Not to eat, you understand, just to regard. Tenrecs are mammals best described by what they're not: they aren't shrews, platypuses, hedgehogs or moles, but have something in common with all of these. I couldn't wait to see one.

Right then, though, we were heading up a trail to Amber Mountain. The forest was dark, dense and full of things with names like dragon trees, flaming katys, polka-dot plants and outrageous, balletic orchids with tongue-twister titles like *Aerangis*, *Jumellea bulbophyllum* and *Phaius*. It all seemed to belong to some remote epoch. In this landscape it would not entirely surprise you to see pterosaurs gliding through the trees and velociraptors sprinting across the trail ahead. But the more we hunted for tenrecs, the more they weren't there.

'They're not *fady*,' Zak explained, after another fruitless forage in the soggy leaf litter. 'Lemurs are *fady*, so are chameleons, but not tenrecs. So they get eaten.'

We had time on our hands, maybe another ten hours of hard climbing, so it seemed okay to begin yet another conversation about the complexities of Malagasy customs.

'What's *fady*?' I asked.

Zak, you must understand, is not your common sort of guide. His father was a musician fairly famous in northern Madagascar, but he died when Zak was quite young. Zak and his six sisters were brought up in a peasant village by his mother and beloved grandfather, who was both a champion bare-knuckle boxer and a storyteller.

In his youth the old man had been press-ganged by the French into building the road to the top of Amber Mountain upon which we were walking – though after fifty years of neglect it was a mere precipice-hugging, tangle-foot path.

Zak learned French, then English, then Italian, and decided growing rice and mangoes wasn't for him. His ambition led him eventually to York Pareik, who runs a travel outfit named King de la Piste. A *piste*, in case your French is as lousy as mine, is a dirt road, or a track, or a ski run ...

Zak soon learned the Latin names of almost every Madagascan plant and creature. So now – in his late 20s, with a head full of tribal customs, hundreds (maybe thousands) of traditional stories and a fine grasp of modern ecology – he's a sort of Renaissance man. You ask him a question and you invariably get a very full answer plus peripheral anecdotes.

To radically paraphrase his answer, *fady* is a taboo system so complex that neighbouring villages and even close neighbours don't necessarily share it. Taboos can vary from community to community or family to family, even person to person.

Perhaps eating pork is *fady*, or digging a grave with a spade which does not have a loose handle is *fady* (not too much contact between the living and the dead). In Imerina area it's *fady* to hand an egg to someone; it must first be put on the ground. In many areas it's fady to work in the rice fields on Thursdays, or work at all on Tuesdays.

Places are also *fady*, and all over Madagascar you will see trees or rocks lovingly cloaked in bright cloth, or bowls full of money beside certain objects. We even came across a skeleton in a cave with a platter of coins beside its grinning skull.

A close relative of *fady* is *vintana*, which cuts up time into good times and bad times to do things, which means people might suddenly stop what they're doing and sit down for an hour or two.

Some things, though, are generally agreed to be *fady*, like chameleons and lemurs. Chameleons, from tiny scraps of rainbow to huge, near-metre-long dragons, are everywhere and, in the forests, lemurs hurtle all over the place – or sit and stare at you. Some will even steal your lunch.

It's just a pity there's no taboo on slaughtering rainforests – only about 10 per cent of them are left and vast areas of the island are either eroded ruination or rice paddies. Well, there you are.

Now to the purpose of my trip: I'd come to Madagascar to explore two mountains, one entirely cloaked by some of the surviving rainforests, the other ripped to shreds by water. I was filled, like Kipling, with a 'satiable curiosity about an island where almost everything was endemic, volcanic or just plain odd'.

Any trip necessarily begins in Antananarivo because that's where international flights land. It didn't take me long to realise it wasn't Africa. Antananarivo is a traffic jam clamped round a curious, pointy-roofed city surrounded by endless rice paddies. It takes a while to get anywhere, but the people along the roads are charming and mostly beautiful Indo-Malayan or Malayo-Polynesian, so it's not a chore to grind along in first gear. Coming in from the airport, I just sat and gawped at all the delightful strangeness.

But my mountains – Ankarana and Amber Mountain – were in northern Madagascar, so I flew out next day to Antsiranana and King's Lodge. York, who owns the lodge, is seriously laid back. He has a fleet of elderly but serviceable Range Rovers plus some year-old twins who keep him and his wife, Lydia, awake at night. There was a time, he confessed as we clutched cold Three Horses beers and watched the afternoon sun get swallowed by a volcano, when he wore his hair long, sampled interesting substances and travelled the world in dangerous public transport. Then he found Antsiranana and decided he'd arrived at his Shangri-La.

I couldn't argue with him about that.

Next day Zak packed up our kit and victuals, picked up a shy cook named Bridget and set off on the grimiest journey I have ever experienced – and I work in Africa.

Madagascar's not dubbed the red island for nothing. When forests are slashed and burned, the oxidised laterite soil beneath them is singularly infertile and soon erodes, spreading fine ruddy powder over the whole island. In the rainy season roads are a quagmire; when they're dry each bump produces an effect not unlike hitting a fat bag of bright red flour.

When we arrived at the spiky buttresses of Ankarana we were astonishingly red, with our personal colour showing only where dark glasses had been. Each rivulet of sweat made the mess even more surreal and sticky.

Well, never mind all that. Let me tell you about Ankarana. On an island of strange things it has to be near the top of the strangeness list. It's a limestone massif sticking abruptly out of the plain and covered in razor-sharp limestone karst spikes known locally as *tsingy*. Instead of going round it, rivers chose to run through it, forming caves, canyons and eerie underground tunnels. You can walk through several hundred *kilometres* of these things. In places huge caves have collapsed, forming isolated pockets of river-fed forest with sunlight streaming down from above. Creatures live in there you'll find nowhere else on the planet.

We hiked through the inky black caves, some with stalactites and stalagmites twinkling in our torchlight, then climbed up the mountain to view the forests and rivers from above. A troop of rare crowned lemurs came to investigate. I bent down to photograph one, and three others jumped on my back to see what I was doing. *Fady* works wonders in the trust department.

We camped at the foot of the *tsingy* mountain, watched the setting sun turn them orange, and Zak produced some local rum. It was spittingly awful. When I refused to sip more, Zak told me a story which seemed to justify getting motherless on bad booze and opened up a later discussion about why the Malagasy are a nation of grave diggers.

There was once a *mamalava* (rum drunkard) whom nobody respected. He was around during a *famadihana*, a time of year when, for some reason or other, corpses are exhumed, their bones dusted off, danced with and carried to some other place (I saw empty graves all over).

When the skull of the skeleton being exhumed was exposed, it began to move this way and that. Everyone ran away in fright. But the *mamalava* took a closer look and saw it was being moved by a little tenrec in the brain cavity. He called everyone back and after that they all approved of his rum habits.

'And that', said Zak triumphantly, 'is why it's good to drink Malagasy rum!'

Well, maybe.

Next day we headed through more red dust to Amber Mountain. It's a thing apart, a great volcanic massif covered in ancient mist forest with its own wet microclimate. Manokan Ambre, as it's called, towers over the northern tip of Madagascar with, more often than not, a soggy cloud frown across its lofty peak.

We overnighted in a very neat, clean hut at Roussettes Forestry Station and hit the trail in a light drizzle at an ungodly 05h00 the next morning. Half an hour after passing an atmospheric little waterfall (sacred, of course) near the hut we were gazing down into the forest-cloaked mouth of a crater lake named Mahasarika.

The forest was … weird. A glance at the statistics will tell you why. Madagascar has around 10,000 plant species, of which about 80 per cent are endemic. It has *eight* species of baobab whereas the whole of Africa has only one. Of the 258 species of birds, 107 are endemic. Then there are lemurs – 30 species. They're primates which look like monkeys with foxy faces and cat eyes. And tenrecs, which Zak promised …

We hiked past huge panda trees *(Pandanus)*, fluffy-topped Araliaceae literally dripping with epiphytes, many of them orchids, huge manaries *(Dalbergia)*, massive tree ferns *(Cyathea)* and spooky dragon trees *(Dracaena)*.

A turkey-sized Madagascar crested ibis, the bane of tenrecs and chameleons, scratched in the understorey, a monticole played hide-and-seek in the upper branches of a manary, and Madagascar bee-eaters buzzed through the mist-edged glades. Lurid lichens sprouted from the rotting trunks of trees thrown down by the area's violent cyclones and in the stream beds boophis frogs muttered contentedly.

The place was simply glorious and, as far as humans go, we had it all to ourselves.

Around five hours deep we came upon a moody crater lake which appeared suddenly as we stepped out of the forest onto its grassy apron. From there it was a slippery, tough climb out of the crater and up to the summit, marked by a neat concrete pillar and a plaque with the words 'Pic d'Ambre'. My GPS registered an elevation of 1,488 metres and, in case you like exact locations, it read S12°35.778, E43°09.188. In the swirling mist we could not see more than a few metres.

Nearly an hour later, after we'd munched the scrambled egg sandwiches and fruit Bridget had prepared, the murk suddenly lifted. Below us was a great crater – I suspect it was the one named Renard but we weren't sure – and beyond that the Mozambique Channel. There's something about the view from the highest point of anywhere that makes it worth the pain.

Twelve hours after starting we were back in the hut. My feet hurt and I was suffering from the strange emptiness I often feel after climbing a mountain. Maybe I looked glum.

Zak, ever perceptive, handed me a glass of rum and said: 'Would you like me to tell you a story?' He gave me no time to reply.

'There were once three friends, *bon*? A duck, a chicken and a dog. They lived together in a village. Once every week they'd take the *taxi-brousse* to the market in Antsiranana. But one week, when they were half-way to market, the taxi driver he told them that because of the cost of *gaz-oelie*, the fare had gone up from 300 francs to 500 francs each.

'The duck he had the extra cash, the dog was rich anyway, but the chicken he had only 300 francs. So when they arrived in the market the duck got off and paid 500 francs. The chicken paid 300 and told the driver that his friend, the dog, would pay the rest. The dog had a 1,000-franc note. The driver took it, got in the *taxi-brousse* and just drove off.

'That is why, today, when you see a duck in the road it's not worried about the car, the chicken always runs away shouting and flapping, and the dog he chases the car going "woa, woa woa" …'

'Are you feeling okay now?' Zak asked, looking at me with his wise, serious eyes all etched about with smile wrinkles.

'Yes, sure. Fierce rum, though. Do you think if I drank it I could spot a tenrec?'

'Well, we could try.'

Ballets of catastrophe

Water's strange stuff. In a glass, or frozen, or boiled it's reasonably straightforward. But when it starts to flow, the physics of water become unspeakably complicated. It's for good reason that fluid dynamics is one of the most complex branches of physics.

It obsessed Leonardo da Vinci. When he wasn't painting, planning a sculpture, constructing a war machine for Ludovico Sforza, dissecting corpses, designing a helicopter, grinding lenses or planning some great building, he could be found peering into rivers.

He'd sprinkle grass seeds into flowing water to see what happened to them in whirlpools. With superhuman quickness of eye, he'd sketch patterns in the near-infinite complexity of fluid vortices.

Crammed in an over-small kayak and heading for a gargling

vortex which kayakers call a hole, my thoughts were far less lofty.

Good winter rains had fallen the week before and that night it would snow. One can become lyrical about the beauty of the scenery around Kleinmond, but right then the Palmiet River was simply a ditch sluicing countless tons of water off the mountainside. The tea-brown stuff was being whipped into an ugly froth, hurtling over, round and under obstacles, going left, right, downstream and even upstream, forming high-speed vees, holes, strainers, rooster tails, standing waves, surges, pillow waves, eddies, aerated bubble banks and nasty barrel rolls known, appropriately, as washing machines.

We'd spent part of the Saturday being coaxed and chivvied by Andrew Kellett of Gravity River Tours to do things he thought would be necessary to stay alive while hurtling downstream the following day. He'd kicked off with an alarming little lecture on the various ways we'd be likely to die.

So, when I hit the water on the first day, I was filled with images of smashed shinbones, dead kayakers held underwater by 'strainer' branches and – after viewing some photos of human flotsam – an almost hysterical fear of weirs. But what kept me awake that night, anticipating the trip, was ravenous holes.

A hole is essentially a whirlpool laid out on its side, with its axis of rotation perpendicular to the main current. It's a cylinder of water and froth that recirculates constantly, in position, like one of those giant spinning brushes in a car wash. For a good approximation of how it feels to drop into one, you could take a pass through a car wash on your bicycle.

Holes, according to an authority on the subject named William Nealy, have to do with vortical hydraulics. The velocity of water dropping over an obstacle is far greater than the water velocity below the hole. This creates an excess of piled water with nowhere to go. Gravity pulls some of it upstream, and the water rolls under, down through the hole again, continuing the cycle.

Experienced whitewater kayakers drop into holes on purpose, using their precarious equilibrium for blatant shows of hotdoggery. A novice who is sucked into a sticky hole does a brief ballet of catastrophe and disappears into the maw. You have an emergency.

Sunday dawned bright and icy. A quick breakfast at the backpacker where I'd worried out the night didn't stop my shivers, but I suspect it had nothing to do with the temperature.

The put-in point was in the Kogelberg Reserve on a deceptively quiet stretch of purple-brown water.

I used some lame excuse about photography to get myself into an inflatable croc. Not much glory there. However, on the first rapid, named Bubble and Squeak, the croc rode the bubbles but failed to squeak past a motor-car-sized rock. It hit the pillow wave side-on, tilted upriver, filled with water and flipped.

Snowmelt water is not the best stuff to fall into. Snowmelt roaring down a series of rapids is worse, and bouncing down backside first is untenable. I'd been parted from the croc and was in mid-stream. I remember thinking: Am I going to do the five kilometres to the bridge like this?

Fortunately, flowing water tends to spit unwelcome objects into eddies, which is where I became reunited with the croc. For some perverse reason the wipe-out had been exhilarating. As I waited, knee-deep, camera in hand, for the others to repeat my performance, my feet turned blue and disappeared as a site of sensation.

Four rapids further I came upon The Waterfall. Here the Palmiet plunges over a ledge into a pool, but there's an escape route of frothing rapids on the left bank. I set up my camera above the fall, focusing on the rapid, and waited for the crowd.

Andrew appeared, gave me a thumbs-up, and began his run. Peering into the viewfinder, I was puzzled when he did not appear on the lip of the rapid. I looked up in time to see him hurtling over the waterfall. His tiny kayak dived like a gannet, did a neat U-turn in the pool and – nose in the air and Andrew back-bracing elegantly – danced across into an eddy and flopped back horizontal.

'Did you get the pic?' Andrew yelled.

'Of course not! Why'd you go over the falls?'

He grinned, hauled his kayak up the cliff and did it all over again. They're a crazy lot, these whitewater types, and they ride crazy boats.

River kayaks were once long, narrow tubes of fibreglass. These days they're spoon-nosed, spoon-tailed, flat, plastic things that have more in common with propellers than boats in which Inuits chase seals. There's good reason for this. Drop a kayak like a Pyranha, Perception or Dagger into a hole and its nose will go down like a shovel, its tail will go skywards and you're in an ender. Lift your right knee and go into a pirouette – if you can – and the game's on.

Viewed from the side, a modern river kayak looks rather like a double blade, so there's not much resistance when you pirouette over your ender. With luck you can come out of your pirouette with a reaching reverse stroke and drop straight back in the hole. Another ender and pirouette, then a cartwheel, a barrel roll, another ender, a full spin, another cartwheel … A novice's nightmare becomes a river-rodeo-rider's delight.

If you want to know if I fell into a hole the answer is: I didn't. I nibbled the edge, heard the sucking and shied away. I have a theory that until about the age of 30 you know you're immortal, after that you know you're not.

I left Andrew, Meyer de Waal and other river cowboys doing endos in a gurgling hole below the bridge. I told myself it was time to go, that I had important things to do, that it was too cold for more kayaking. But I sure as hell would have liked to be in there with them. Leonardo would have stayed. Kayaks have the edge on grass seeds. How else can you watch a river swallow a human – again and again?

Paddling back in time

For some reason the word DANGER surrounded by a red triangle seemed to mark the boundary between the bonhomie of the *boets* and *swaers* of Port Alfred and the heart of darkness. The rain-bruised clouds added to the impression.

Maybe the mid-river sign was merely marking a hidden reef. But its presence had the same unsettling effect that 'Here There Be Monsters' must have had on ancient mariners studying their hard-won maps. Some 180 years ago, when an ancestor of mine set sail for these shores, there were still maps carrying that warning. I eyed the danger sign, wondering about its portent, and paddled on.

Until that morning there had been others happy to kayak up the Kowie River with me. But the day had dawned wet and blustery. They quickly reversed their offers: 'No way! Not in this weather ...' and went back to bed. So my trip up the river and into my own history would be alone.

In a sense, it was a trip with two beginnings. One had to do with a large, maroon motorbike; the other with a sailing ship on a far distant shore.

When I picked up the Kawasaki Vulcan from Yaron Wizman, who rents the things, he looked worried. 'Spray lube on the chain at every petrol stop,' he advised. 'Check the oil and water. Watch out, it's fast.' Maybe I didn't look the motorcycling type.

He was right about fast. But, snaking along the foot of the Overberg in a light drizzle, the rain hammering on my helmet visor with a sound like chips frying, speed wasn't an advantage. The weather was probably only marginally better than in Wiltshire.

Wiltshire, an English county, was the site of the other beginning. Probably in just such a drizzle, a man with my surname sat in a village in the English county and decided he'd had enough of the gloom, lousy pay and a declining demand for swineherds. So he packed his family into a crowded immigrant ship in 1820 and headed for Africa.

Maybe he couldn't read the small print – maybe there wasn't any – but conditions in the Eastern Cape where they landed were less than hospitable. The veld was bitter, there were animals which ate you and the Xhosa were understandably angry at the sudden invasion of pale, plough-toting *abelungus*.

It didn't rain for two years. Then a flood took out the crops. My forebear gave up farming and moved to Grahamstown, where he bought an ox wagon and did transport riding. Every year for generations – right up until my father's childhood – the family would trundle down to the Kowie in a wagon and outspan along the river banks. From time to time the men would brave the monsters upriver at Waters Meeting where they'd hunt for the pot.

These days Waters Meeting is a nature reserve and the best way to get there is by canoe. Conditions there are still, well –

settler simple. There's a hut, a water tank and some bunk beds. All about are wild hills where Afro-montane forest meets subtropical valley bushveld amid symphonies of birdsong. Going there would have all the elements of pilgrimage.

In Port Alfred, when it finally appeared from under the clouds, I found Bev Young, a larger-than-life Harley biker Hell's Angel mama who works in the town's information office and organises everything.

'Hey, Don, you arrived in one piece?' she yelled as I staggered through the door after countless hours in the saddle. 'I must show you the crayfish. Did you know Port Alfred was ringed with game farms? This place will soon be the new Mpumalanga. And there's no malaria. I love this place. Would you like some tea?'

Half an hour later I'd been on a head-spinning verbal tour of the little port: historic houses, fishing, the endless beaches, great surfing, some juicy gossip, 43 Air School, the new marina, freshwater crayfish . . . and cold tea (she forgot to switch on the kettle).

'Oh gosh, I'm sorry,' she chuckled. 'Perhaps you'd better check in to the Halyards and have a bath or something.' I did, then went to explore.

Port Alfred has two elemental features which stand out above all the rest: its beaches and the estuary. I hitched a ride with environmental officer Anton Gouws doing beach patrol up to Three Sisters rocks. The beach must have been half a kilometre wide and the only tracks in the sand belonged to seagulls. The sounds that curled around us as we switched off the engine were their cries and the endless growl of waves. The rocks, sculpted by the hammering sea, looked like three giant chocolate layer-cakes at the wrong end of a toddler's party.

The tidal estuary has another kind of beauty but has been the victim of men's dreams of fame and fortune. The dreaming began with William Cock, who arrived there from Britain in 1836. He built a house on the west bank (it's still there and known as Cock's Castle). Two years later he began diverting flow at the river mouth in an attempt to make it accessible to sailing ships. He managed this task and had a pier built, then went into denial as ship after ship was wrecked crossing the treacherous sand bar at the river mouth.

The port eventually attracted sailing craft – as old photos in the museum attest – but when steam replaced sail, the harbour proved too shallow to take the iron ships. It now serves only ski boats and doughty surfers who leap off the pier into the raging surf.

More recently a glitzy marina was built on Cock's reclaimed land. It's an elegant place with fine houses, but I couldn't help feeling the river would one day spit out the constriction in its mouth and turn the estuary back into the delta it once was. It'd make a mess of all those beautiful gardens.

The upriver reaches, when I went exploring on the bike, were compelling – even though the weather map showed 40 per cent chance of rain with snow on the mountains. Anyway, venturing out in lousy weather ensures you have the world mostly to yourself. As long as you stay warm and dry, it's a traveller's bonus. Rivers Meeting – where I'd spend the night – and the spirit of my ancestors were calling.

Spirits of another sort had called a man named Derek whom I met on the way back into town. He was pushing a huge, box-like contraption on bicycle wheels. It had a propeller on front and messages from Jesus all over it. He'd just bullied it all the way from East London, a few hundred kilometres away.

'We are a forgotten people,' he told me happily. 'But I preach and the Lord keeps the propeller turning.'

Back at the Halyards I consulted Andrew Hutchinson of Hutch's Boats. He took pity on me and offered to tow my kayak up to the caravan park. It wasn't something I was about to refuse, especially as the rain was pelting down just then. I took the hopeful view: by the time the tide began its upriver push all 40 per cent of rain should have fallen. I thought some dark thoughts about my sleeping ex-fellow kayakers and hopped aboard.

Along the way we picked up a food basket from Butler's Riverside Restaurant. It looked way over the top, but I was in a hurry and stashed the boxes into the kayak's storage compartments.

We puttered past the wreck of an old paddle steamer and through the Bay of Biscay (so named because the wind howls across it). At the caravan park Hutch cast me loose. Soon there was nothing but the lapping of water and the liquid piping of a black-headed oriole.

With some help from the paddles, the kayak rode the tidal surge past Cob Hole and I was soon at the ruin of an old mill.

'Would you like some frogs?' asked a man who was standing near it with a face to match the crumbling stonework.

'No thanks.'

'But you need frogs,' he insisted.

'Why?' I yelled back.

'Because fish love to eat them ...'

'But I'm not fishing!'

'Well, you should,' he threw back, then disappeared into the forest.

Soon after that rather surreal exchange the red danger triangle came into view. At a place named Fairy Glen the river turned westwards, narrowed suddenly and became more confidential. Riverine forest dominated by tree euphorbias covered the enclosing hills and, as I turned the bend, a pied kingfisher hovered invitingly. 'Kwik kwik,' it insisted.

Beyond a sharp northward bend the river straightened. Several kilometres ahead was a towering wall of forested hillside.

The valley rang with the cry of fish eagles.

At a spectacular 360-degree oxbow named Horseshoe Bend was a jetty and a sign which read 'Waters Meeting Overnight Hut'. Joseph Conrad would have loved its mysterious invitation into the looming forest. Soon after hauling my gear and food boxes inside, the rest of the predicted 40 per cent of rain descended. I dragged a rough table onto the stoep and unpacked the food.

The contents were so outrageous it was comical: lobster, prawns, meat balls, samoosas, Portuguese rolls with sausages buried inside, fish fingers, cheese, biscuits, apples, bananas, a chocolate cake, a flask of carrot soup, another of hot chocolate, and a bottle of Guardian Peak Cabernet Sauvignon with corkscrew, wine glass and checked tablecloth.

Among the goodies was a note which read: 'Learn to pause or nothing worthwhile will catch up to you.' Was I being spoilt or was this regular Butler's fare? I spread the packages on the cloth, uncorked the wine and tucked in.

A jackal nickered then yipped, fireflies sparkled in the bush and the silence was made somehow deeper by the admonitions of a purple-crested lourie. A small, brown form glided down the path: could it have been the rare barred owl?

Raising my glass, I peered through the ruby wine at the gathering dusk. Just what was the spirit of my ancestors? Would I feel it? How would it manifest itself? Should I invoke it? It seemed a bit childish to call out: 'Hey, spirits of my ancestors, I'm here.' But there was nobody around so I did it anyway.

Moments later a strange moaning began and I felt the hairs on the back of my neck bristling. What happened next nearly undid me. Round the corner, at speed, came a large male chacma baboon. On seeing me a few metres away he let loose a deafening 'baa-hooo' and skidded to a halt. At that moment all the hadedas roosting round about went 'haa-dee-da' together and I found myself standing on the table with the only weapon available: the corkscrew.

The baboon bounded off, the birds quietened down, and in the absolute silence that followed I could hear my heart thumping like a tom-tom.

The rest of the night was quiet and the only visitor was a wily, larger-spotted civet, which arrived to polish off my leftover prawns. Dawn was glorious, with rays of golden sunlight levering the remains of yesterday's clouds out of the way. I packed up and set off down river on a falling tide.

With three hours of paddling ahead, I had time for reflection. It struck me that I should, at least, consign to print this message to my descendants and anyone else who may be interested: Beware of invoking the spirit of your ancestors. You never know in what form it may appear.

Travelling the trails of !Xam

It was a matter which could have been cleared up by a single conversation. But few settlers realise they owe it to history not to shoot first and ask questions afterwards . . .

I sat gnawing at the problems caused by the short-sightedness of my ancestors, staring rather sadly at an extraordinary scene. A charging lion, about half a metre long, its whiskers bristling and its teeth bared, was painted on the cave wall with the definite strokes of a master artist.

Ahead of the beast, seeming to flee with great leaping steps, was a row of San hunters. Above them floated little red boxes which resolved, Esher style, into flying figures with antelope heads. These spiralled up the domed roof and into a circular indentation at its highest point.

The small cave I'd ducked into was an almost perfect dome, some two metres at its highest point and maybe five in diameter at its base. We'd found it atop a boulder-strewn hill behind a gracious stone farmhouse in a valley named Balloch, part of an area in the Southern Drakensberg known, rather oddly, as Wartrail.

The tableau before us was dream-like, vivid; teasing cognition but inexplicable. The cave itself had a strange, alien presence and I found myself glancing out of the opening expecting someone else to arrive. But no one did. The !Xam had long gone: hunted down, shot, starved or frozen to death in their mountain hideaways.

At the turn of the century the historian George McCall Theal, with typical Victorian arrogance, had put the matter quite plainly. Bushmen, he said – using a name which was to be much contested later – 'were of no benefit to any other section of the human family. They were incapable of improvement, and as it was impossible for civilised men to live on the same sod with them, it was for the world's good that they should make room for a higher race.' Presumably his own.

More recent historians would call the events which Theal supported by a more uncomfortable name: genocide. The San were forced to make way, and sitting there I realised there was nobody on earth who could now tell me with certainty what the painting on the cave wall meant. There are theories – books full of them – but San rock art still awaits its Rosetta Stone.

Some months earlier I'd phoned Susan Tonkin of Wild Cape Ventures in Ugie to ask if she could suggest a route that would take in the best San art sites of the Southern Drakensberg.

'It's possible,' she chuckled. 'But you'll need a 4x4 and some good hiking boots. They can be wild mountains.'

So September found me thrumming my way through Queenstown and Dordrecht to Ugie in a beefy 4x4, wondering if I was in for baking heat or ice: these bergs can lay down ski-depth snow on Christmas Day and fry you in July.

Ugie was balmy, with little white clouds giving no hint of snowstorms. I picked up Susan (we'd soon rename her Su San) and her forester friend Gordon McKenzie and headed up to Woodcliffe where Phill Sephton owns a farm so outrageously beautiful it hardly seemed real. Beyond the lawns of her self-catering cottage massive, muscular sandstone cliffs rose almost vertically out of the river and continued doing so up the valley until they lost themselves in the purple basalt stickleback of the Drakensberg.

Next morning we followed the river upstream, past fields where rare crowned cranes gathered, then skirted valleys of indigenous riverine forest boasting Outeniqua and real yellow-woods *(Podocarpus sp.)*, white stinkwoods *(Celtis sp.)*, cheese-woods *(Pittosporum viridiflorum)*, horsewoods *(Clausena anisata)* and Cape quince trees *(Cryptocarya woodii)* among many others.

The first cave offered a beautifully drawn cheetah, the second a single baboon. Then we trekked out of the river bed and onto a ridge overlooking Wide Valley with a splendid view of the high Berg. In a most unexpected place Phyll pointed out a rain-animal painting that looked to me like a nasty-tempered moray eel. Near it was the first of many 'therianthropic' figures we were to see – humans in the process of transforming themselves into animals. Around it were masses of dots and lines known as entoptics: patterns seen in a trance state.

It doesn't take much stamping round San caves to realise their paintings weren't 'art in the park'. Frankly, they're weird. In the absence of a San shaman to explain them, the next best interpreter is Professor David Lewis-Williams of Wits University's Rock Art Institute.

In 1990 he published a useful little book entitled *Discovering Southern African Rock Art*, which was followed by the much more weighty *Images of Power*, then *Fragile Heritage*. Most of the images, he suggests, were painted by shamans re-creating their own spiritual experiences, and he has collected a mountain of evidence to support this view.

For the San, it appears, there were two worlds: that of the camp and surrounding wild beasts, and that of the world associated with the supernatural and with strange creatures – a place of the gods from which power could be drawn. Between them were what could best be described as portals: certain shimmering water holes and cracks, or special surfaces in quartzitic rock. Doorways between the two worlds.

The intermediaries between these worlds were shamans who, through trance-inducing dance rituals, could 'die' in this world and travel through portals into the spirit realm, seeking the power to heal, make predictions or bring rain.

From dances witnessed among living San, and from the perilously few records available, it seems that entering a trance could be painful, causing the shaman to double over and bleed from the nose. Often they would take on a therianthropic form, 'becoming' an animal – generally an eland. Often these creatures would be bleeding from their noses. What they experienced they painted: trail maps to other worlds.

By the time we returned to Woodcliffe cottage we'd covered some 13 steep kilometres – a tough start to a week in the mountains.

Next morning we pointed the 4x4's nose towards the picturesque village of Rhodes. First, though, we dropped in on !Kaggen's Cave and its fading herd of eland, then round on the contour to Outlook Cave. Sitting there, doing what the name implied, a deep, healing stillness seemed to enfold us as we stared across at the emerald foothills of the ever-present Drakensberg. The next site, Storybook Cave, had two metre-high therianthropes – one pointing authoritatively at a crack in the rock from which an eland emerged – and a two-tailed cat.

It was about that time I realised San rock art was not so much about pictures but about place. The caves and overhangs were rather like theatre proscenium arches marked 'entry', surrounded by elaborate scripts on how to enter and about what to do beyond the portals.

This trip was becoming a tour of the doorways to San heaven. But where on earth were the keys?

The drive which followed up Naude's Nek Pass was one of those which people buy 4x4s to experience, though we got up with only the back wheels churning. One false move and we'd have become a panelbeater's nightmare. Once over the top the vehicle was invaded by the delicious aroma of blossoming thickets of *ouhoud (Lecosidia sericia)*, which seemed to grow everywhere.

Rhodes is one of those villages which time forgot before newness became a fashion. Nestled in a valley of the Drakensberg foothills, its houses have deep verandas, steep corrugated-iron roofs and sagging wire fences.

We zigzagged down the wide dirt streets, bemused by the almost over-abundance of rustic charm, then pulled into Walkerbouts Inn. It's owned by Dave Walker – self-proclaimed mayor and Mr Trout – and isn't short of either rustic or charm.

The bar counter was made of solid cedar and the fish tank had recently been vacated by some trout because it had sprung a leak. They languished with a platanna in the rather cramped quarters of a cooler bag with a bubble machine attached.

Dave's a large, easy-going host who came to Rhodes because of trout. He organises fly-fishing trips, runs the inn and fires up a mean pizza. Up the road is the ski lodge Tiffindell, but he's quite glad the favoured road to it bypasses the village.

'The yuppies go up there in their fancy 4x4s. Here we get interesting travellers and sheep farmers. Good people.'

Just outside Rhodes the next day, Vasie Murray – farm owner and sometime film-set animal handler – took us up a valley which loomed in on the road and made each corner seem sinister. Desolate beauty might be a bit of a cliché but it's the only phrase which seemed to fit.

Martin's Dell Cave, high up one of the valley sides, had eland paintings so highly coloured they looked as if their long-gone painters touched them up annually. Above one was a white bird doing a high-speed dive into the back of a staggering grey eland.

In another cave, once filled with art, a farmer who had used it as a shearing shed painted the walls with whitewash to improve the light. In cultural terms it simply extended the gloom.

Nearby, Willem Naude of Buttermead Farm – a man with a passion for rock art – showed us a cave with a painting of what has come to be called the lightning bird. It was connected to the bleeding nose of an eland by a zigzag line, possibly representing supernatural potency. If these little artists were primitive they certainly used some very sophisticated metaphors.

A mountain buttress and many kilometres later we followed the directions of Alan Isted to Bidstone Guest Farm, which is run by his parents. His mum, Di, cooked up a fine meal and we sat clutching beers and discussing the idiot weather. Alan's a snowboarding man and was chafing at the bit. It was hot when it ought to be snowing: global warming, undoubtedly, we agreed. In a few months time he'd be off to the Himalayas for the real stuff.

Alan's knowledge of rock-art sites is considerable – he's been hunting them down much of his life. The first place he took us to was Warwick's Cave at Balloch – the domed cave of the lion chase. The second required some serious walking through high mountain canyons to Brummer's Cave.

By this stage it might have been the daily hiking, the bizarre paintings or continuously hunkering down in places of power, but the trip was getting increasingly surreal. Brummer's Cave would blow me away completely.

The place was as large as a big concert hall, with a view over the Kraai River valley, and was simply full of polychrome elands. They were drawn with such clarity and understanding of tone they seemed three-dimensional. Between them were busy human forms, curled snakes and other, odder creatures.

The paintings were perfectly preserved, possibly because they're so deep inside and far from weather and water erosion, but also because they were on two levels, both protected from cattle and sheep, which obviously use the cave. And they're certainly way off the tourist map.

The looming buttresses beyond the vast cave mouth and the valley far below offered no hint of human life, but behind me the walls vibrated with evidence that this was once a San equivalent of France's Lascaux, and possibly far older. The presence of the little hunters was so strong, the silence which enveloped us seemed but a gossamer portal away from their clicking chatter.

These Drakensberg foothills had been home to the San for countless centuries, some moving with the seasons and others settling permanently in great caves such as this one.

When the Nguni cattle herders appeared a thousand or so years ago the San simply moved to higher ground and – because they were so few, had no cattle and posed no threat – they were left in peace.

But from 1837 Dutch farmers, moving away from British rule, trekked into Natal and before long were in conflict with the black pastoralists.

After the defeat of the Zulus at the Battle of Blood River, settlement in the region increased. Wild game was quickly depleted and the San, deprived of their food source, raided livestock. Farmers retaliated with vengeance, shooting San 'pests' on sight. When Britain took over Natal, English settlers simply kept up the tradition.

While the Natal Volksraad hadn't exactly specified extermination, the instructions were so broadly worded that discretion in that matter was left to the local *commandant*. Thousands of San died. With a total estimated population of around 20,000 for the whole of South Africa, the effect of the virtual open hunting season on the San was devastating.

The final demise of the southern San is chillingly depicted by historian Nigel Penn: 'In the desolate obscurity of the 19th century *agterveld*, the San were overcome by a piecemeal process of betrayal and defeat. By the 1870s the last remnants of the Cape San were being hunted to extinction. Those who were not shot were starved to death in the dusty margins of South Africa's most marginal land … '

That sad history seemed oddly out of step with the elegant images behind me and the breathless beauty rolling in from the yawning cave mouth. I guess a conscience is what hurts when everything else feels so good …

We left the cave reluctantly, and as we crested the valley a howling gale stopped us in our tracks. Driving wind saps the spirit, which may have explained why we arrived back at our vehicle feeling rather flat and disgruntled. Or had we disturbed something ancient?

From Wartrail, we drove through the historic but unattractive town of Aliwal North and decided to push on to overnight at its aesthetic opposite: Burgersdorp.

After a fine meal and a comfortable night's rest at The Nook B&B in the care of Anita Joubert, I took a dawn stroll to investigate the Anglo-Boer War blockhouse. Along the way I had the unusual pleasure of being simultaneously crowed at by a rooster and barked at by a crow in someone's garden. When a tough-looking Staffy came up, yapping to get in on the act, the obviously tame crow beat him up and sent him packing.

The rock-art trail would end some days later at Greenvale Cave in Dordrecht, a thousand kilometres from where it began.

The cave has a strangely lyrical 'flute player' and some freakish nightmare creatures. But, in a sense, my personal quest for the spirit of the San ended in a canyon near Burgersdorp so remote even the 4x4 seemed nervous – and so full of exquisite paintings it should immediately be declared a national monument and a World Heritage Site.

Known locally as the Valley of Art, it spans the farms of A C de Klerk and Ouboet Coetzee. They're well aware of its importance and are dedicated to its preservation. But sheep farming has fallen on hard times and there's no guarantee their properties will remain in sympathetic hands. Many farms in the area have been abandoned to weather – and probably the Land Bank – and are sliding into ruin.

Even at the dry end of a particularly dry season, golden-hued streams ran through rustling reed beds and slid into deep pools cradled between towering cliffs of orange and black sandstone.

Almost every overhang seemed to be an art gallery teeming with images of profound sophistication. In one a group of hippos clustered in near-photographic perfection. In the Cave of Birth, amid a welter of polychrome antelope, therianthropes and unintelligible symbols and dots, was the drawing of a woman which was so graphic it must rate as one of humankind's earliest pornographic works. In Rainmaker's Cave strange, bloated creatures loomed while busy little figures towed them magically to bring an end to the dry months of winter.

But it was in the Cave of Dogs that I stared, dumbstruck, into the joyous soul of a departed people. Possibly because it's south-facing and on less friable rock, the images have been preserved down to the feather-strokes in the headdress of a dancing shaman.

The wall was alive with hundreds of figures – here a hunting group with dogs, there a family group on the move; the men with bows and spears, the women cloaked with karosses, supply sacks thrown across their shoulders and digging sticks in their hands. Tall figures with antelope heads strode beside bent, old people leaning heavily on sticks. Marching with the throng were sheep (the San kept sheep?), prancing elands and packs of loping hounds. Three of the dogs were no longer than a centimetre each, but so perfectly drawn that you could feel the joy of their gambolling.

This is how almost *all* the caves we'd seen must once have looked. What were they telling us? The figures were all walking, running, striding across the huge tableaux: all moving. But to where? Whatever destination they had in mind, the real answer was terrible: to oblivion.

As I stood, staring in wonder, my eyes filled with tears of shame at what had been done to their culture. When some *commandant* pumped a bullet into the breast of the last San artist, he would not have known that he'd murdered Africa's equivalent of Leonardo da Vinci or Renoir. Would he have even cared?

How much richer the world would have been if, instead, they had stood side by side before one of the great panels while the artist interpreted his images. But they didn't, and the meaning of San art remains tantalisingly just out of reach. That makes it so intriguing, but also ineffably sad, like a dream you knew would change your life but which you forgot on waking.

History documents so many misunderstandings, but so few meaningful conversations.

Above the sea of mountains

My eyes registered the gradient as gentle but, after 35 kilometres at around 3,500 metres above sea level, my legs calibrated each step in degrees of pain.

The deeply grooved pony track led up through giant lobelia *(Lobelia rhynchopetalum)* and huge erica trees *(Erica arboreum)* towards a sky which swirled and rumbled, threatening rain. I found myself chanting my usual mountain mantra: 'Why, why, why do I do this?'

My goal was Imet Gogo (The Mother), a lofty eyrie on the rim of the Great Rift Valley in north-western Ethiopia – but that was still a long way ahead.

Inevitably, being Ethiopia, two youngsters appeared, seemingly from nowhere.

They were dressed in blankets slung, burnous-style, round their bodies and over their heads. One carried some dropping-spattered chicken eggs in a goblet-shaped basket made of tightly woven sisal.

'Eggs?' he said, hopefully. When I failed to respond the other one tried another tack:

'Hellowhatyournamedoyouhavepenforme?'

'What?' I croaked at him, then realised it was a heavily accented version of a standard local greeting to foreigners. But I was too far gone to be polite: 'Look,' I replied, knowing he wouldn't understand a word I said. 'I've just walked all day through these crazy mountains of yours. Two of my party are back there in that erica forest, suffering from exhaustion and altitude sickness. Now you want me to give you a pen and buy your damn eggs.'

The two looked at me round-eyed, obviously impressed by my speech. Then, simultaneously, they said '*ishee*' (cool) and continued dogging my steps up to Gich Camp. It was, I grumbled to myself, one of those situations in which the idea of hiking the Simiens had been too thrilling to allow better judgement to prevail – and now it was too late to pull out.

Earlier that day fellow traveller Neil Lee had done the sensible thing. When a Toyota Land Cruiser had appeared magically at Sankaber he negotiated a price with the driver and slung his pack in the back. His face at the window, as the sturdy vehicle headed back to Gondor, looking both disappointed and relieved. I should have heeded the warning signals back at Debark. In Amaric its name means, appropriately, 'not fair'. It certainly wasn't: it seemed to be filled with the worst sharks I'd ever encountered offering for hire – at hugely inflated prices – dirty camping mattresses, used maps of the Simien Mountains, and lifts to Sankaber at costs that would be the envy of Johannesburg's taxi drivers.

We had trudged up through the village's market place ahead of the pack-mules, hoping to escape the incessant peddling of kit and services. Apart from some corrugated-iron roofs the place could have been in any century. Hundreds of hopefuls, strewn all about, purveyed richly coloured cloth, startlingly red peppers, huddled sheep, bolts of bright cloth, Ge'ez Bibles, beads and an impossible assortment of junk I couldn't imagine anyone buying.

After hours of slogging through uninspiring peasant fields and dongas, we had pitched camp at Mindigebsa in a downpour.

'It never rains in Ethiopia in October,' our guide, Bedassa Jote, had insisted, looking offended as he coaxed a flame out of the Primus while a drenched mule-minder held an umbrella over him. He had soon produced strong, hot tea followed by pasta and a fresh tomato-and-garlic sauce which we had gulped down thankfully before retiring to bed at 19h30.

The trip to Sankaber the next day was peasant-picturesque but hadn't felt like we were in a national park: it was a non-stop tableau of Old Testament, robed people riding donkeys, shepherding sheep or ploughing steep hillsides with tiny oxen and primitive ploughs.

The Simiens had made themselves felt at the Lamma River, where the path led up and up to a higher plateau. Sankaber had been nothing to write home about – a few huts and a wall-less roof for visitors to shelter under. From here a road had swept us three remaining travellers plus guide, loaded mules, mule-minders and an AK47-toting warden down into a natural meadow which formed the head of a yawning valley offering tantalising glimpses of the Great Rift Valley.

But at that altitude the day's hike had been too taxing for us. Anita Arnot and Ann Griffoen had sat down suddenly somewhere on an endless slope and I guessed it to be altitude sickness: I hoped it wasn't something worse. All that was left of our party was me and my aching legs – and now my companions were two crazy kids trying to sell eggs. If someone got really ill or injured, help was a long way off.

The mules had gone ahead to Gich Camp, the last overnight stop before Imet Gogo, and the little blue tents, when they appeared over a rise, were as welcome as the Addis Hilton. Two mules were quickly dispatched to save the weary hikers and, after a rehydration process and a mug of whisky, the insanity of being here subsided a little. A bright Venus dragged the Milky Way into view, the fire made our sodden boots steam, and rice alfresco and roasted maize appeared. In the distance a Simien wolf yipped.

'Not bad,' I thought, but my mind refused to entirely justify being perched on a flea-infested saddle blanket 3,500 metres above sea level. This was big, wild, hard country and we weren't at the top yet.

Next morning our party was reduced to two plus guide, muleteer and warder; Ann having opted to spend the day at Gich Camp. Anita wisely commandeered a horse and attendant peasant and set off for Imet Gogo in style. The trail led up through forests of giant lobelias with strange flowers soaring up to eight metres above their aloe-like leaves.

After about an hour and a half of steady uphill, our peripheral views seemed to narrow – as they sometimes do on high mountains. As we walked out onto Imet Gogo's rocky shoulder the trail vanished into nothingness and silence.

To the south a massive gorge, with sheer cliff walls higher than Table Mountain, sliced our promontory from the next finger on which we could see the tiny huts of Chennek Camp. In the gorge waterfalls plunged like silent lace, funnelling water into a tracery of streams and rivers two kilometres below us.

Further south, another finger probed the abyss: Bwahit, at 4,430 the second-highest peak in the Simiens. Beyond it loomed Ras Dejen, 4,543 metres and the fifth-highest peak in Africa. Clouds hovered protectively above it and glaciers drew a white line beneath its crown. Below us, to the east and north as far as the eye could see, was Dip Bahir Wereda – the Sea of Mountains.

I sat down suddenly, tears welling in my eyes, such was the raw beauty which lay just beyond my boots. The empty silence was broken by the hiss of wind over feathers and a huge lammergeier, catching an updraft, wooshed metres above our heads. It was so close I could see its golden eyes and the subtle adjustments of its primary feathers. Far below, a rare walia ibex with massive horns hugged the base of a precipitous cliff.

I found myself chanting my other mountain mantra: 'Yes, yes, yes ...'

In this area some 40 million years ago great cracks in the continent's crystalline bedrock allowed the basaltic larva beneath to ooze up and spread over an area covering nearly a million square kilometres and, in the Ethiopian Highlands, reaching a height of more than 5,000 metres. About 20 million years later Africa's crust parted, forming the Great Rift Valley – a 6,400-kilometre-long rift stretching from Mozambique to Jordan in the Middle East.

As the Rift Valley opened up, the larval cap was literally torn down the middle. Glaciers, then water, ground and furrowed the crack ever wider, separating the Simien and Bale mountains, forming the yawning chasm stretching away before me and creating some of the most spectacular highland topography in Africa.

I plodded back down from Imet Gogo in a triumphal procession consisting of several hundred gelada 'lion monkeys'. These large, bewhiskered primates – also known as bleeding-heart baboons because of a red, heart-shaped piece of exposed skin on their chests – were communicating in sounds so close to human speech it was eerie. They were unafraid of us or our horse as they foraged and chatted, merely turning their backs on us to indicate displeasure if we approached them too closely.

The males of this species, once widespread throughout Africa, really do look like lions and, when they bare their teeth, are a fearsome sight.

The two-day return trip to Debark was exhilarating – our legs had been toned by the upward trip, our lungs acclimatised to the altitude and the trail was mostly downhill. Also, somehow, we noticed more going downhill – possibly because the landscape is below the level of one's gaze and not above it.

The mountain flora was strangely familiar: it's basically giant fynbos. Evidently, many of the plants which now constitute the Cape Floral Kingdom began in these highlands and, over millions of years, migrated down the Rift Valley mountains and the Drakensberg chain until they could go no further south. As good travellers should, they reduced their bulk along the way, but in the Simiens, forests of erica trees jostle with giant geraniums and everlastings. St John's wort *(Hypericum)* also started out from the Ethiopian Highlands but only got as far south as Mpumalanga, where its common name is curry bush There were few birds at these altitudes, but the valleys rang with the deep-throated *ha de haa* of the endemic wattled ibis – this species has a gravelly call that makes hadedas (which are also found in Ethiopia) sound like hysterical schoolgirls. And whenever we ate, we were seldom without an attendant cluster of thick-billed ravens. They thumped down heavily, almost within arm's reach, their huge, wicked-looking beaks keeping us on the alert.

As we descended, peasant cultivation brought a marked change to the scenery. It was difficult to believe that people starved in Ethiopia. All about were fields of teff, a grain used to make *injera* – a kind of pancake and the country's staple food. Between them were tracts of wheat and beans. Rivers flowed strongly down the many valleys and plunged thunderously over waterfalls. Although the east and north-east of the country is dry and infertile, the bounty around us seemed capable of feeding the nation for years. I'd hazard a guess that two things prevented this: there are no roads to get the produce out, and peasants, who constitute most of the population, are notoriously difficult people from whom to extract a surplus.

As we tramped down the road out of the mountains a beautiful little girl with tight braids fell into step with me. I estimated her age at about six. She looked up at me shyly, her large eyes twinkling above the fold of her cloak, and began the standard greeting: 'Hellowhat'syournamehaveyougotpenforme?'

I dug in the bottom of my camera bag and produced an old Bic. She clutched it with both hands and gave a little hop-skip of delight.

'*Ishee, ishee* ...'

'Yeah, sure, *ishee.*'

'*Birr* (money)?'

We were definitely back on the road to Debark.

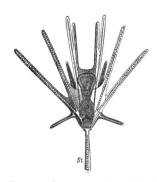

Into the sands of silence

No bird greeted the dawn. There was not a leaf to shimmer in the slanting rays, no breeze to blow it, no insect to chirr, no sound at all.

As the yellow light probed between sun-blackened, time-sculpted rocks of the Akakus Mountains it set the peach-pink sand aglow. When I adjusted my position, a small rivulet of sand skidded down the dune face, hissing loudly in my silence-shocked ears.

If you placed your finger on a map of Africa a bit north of the midpoint between the coasts of Libya and Nigeria, Mauritania and Eritrea, you'd be pointing to the spot where I awaited the sun: right in the middle of the Sahara. It was a very strange place to be. It was an even stranger place to begin a hunt for savanna giraffe, ibex and elephant. And Bushmen.

But when I returned to camp after listening to the desert's

unnerving stillness, the camels were well rested and the Taureg guides had breakfast simmering. It would soon be time to travel and, hopefully, begin to unravel a thread in the history of Africa's most mysterious people.

The Sahara had not always been this empty. A growing body of research suggests that some 6,000 years ago – a mere yesterday in geological time – it was savanna fed by long, meandering rivers. Petrified forest vegetation has been unearthed. In the Murzuq Sand Sea of south-western Libya the bones of crocodiles, hippos, elephants and antelope have been found.

Until the great desiccation of North Africa, caused by seasonal shifts which are still only partly understood, giraffes munched on acacia trees along now long-dead rivers. And in rock overhangs Neolithic artists ground pigments and painted the animals they hunted and the strange, other-worldly visions induced by trance dances. The oldest bones yet discovered in the Sahara belonged not to Negroid or Arab people but, astoundingly, to the oldest of all African inhabitants – Bushmen.

You don't have to have a reason to travel to Central Sahara, just a touch of madness. Simply to be there is reason enough. But the possibility that the art was of Neolithic San origin was irresistibly intriguing.

Three months earlier, being in the Sahara had seemed impossible. The United Nations embargo on flights to Libya had only recently been lifted and tourism in that country was embryonic. Visas weren't exactly a problem, they simply took time and had to be okayed by officials in Tripoli.

The first plane booking had to be cancelled because the visas hadn't arrived. Finally we touched down in Libya but, being Ramadan, all flights out were booked up for a month. To get out we'd have to trek to Tunisia and fly from Tunis. Colleague Robyn Daly and I had flown up the Nile, and on to Istanbul, then over a snow-cloaked Mount Olympus to Tripoli. The omens appeared auspicious.

Tripoli, Libya's capital, was founded by the Phoenicians, fell to the Nubians, was built to magnificence by the Romans (who named it Oea), sacked by the Vandals, invaded by Arabs, fell to the Turks, then the Italians – well, you get the picture.

Leaving it was a relief. It has an atmospheric old quarter, some really tasteless newer effigies and insane traffic. It's hectically urban. Less than a hundred kilometres south, though, was the unsettling but exciting presence of the brooding Sahara.

We set the odometer of the Kia minibus to zero and headed for Ghadamès. The air was sharp and the shimmering dome of the sky matched the blue of my backpack. Some 70 kilometres south of the coast the terrain changed to flat scrub desert, relieved occasionally by hardy fig and, oddly, Australian gum trees.

'Camel grazing,' commented our guide, Sherif Shebani of Coast and Desert Tours. 'Nothing much else out here.'

At first the mountains seemed unreal, a mirage balancing on the arrow point of the road ahead. But they gradually resolved into the Western Mountains, Jabal Nafusah, a chain which begins as the Atlas Mountains in Morocco and stretches clear across Algeria, ending in the sea at the ancient Roman city of Leptis Magna, east of Tripoli.

These craggy peaks, rising nearly a kilometre into the sky, are an escarpment which seems to dam the great sand sea. Along this range live the Berber people – forced there by Arab invasions which spread westwards from Egypt some 1,300 years ago and hedged in by the southern desert.

We hairpinned up Jabal Nafusah below the town of Nalut and under the glowering stare of what seemed to be a castle. It turned out to be a ksar – a fourteenth-century grain warehouse and olive press. Looking like the life works of a giant mud wasp, the place was a bewildering warren of grain 'cells' stacked all atop each other and accessed by now-crumbling stairs and ledges. Holes and ducts kept the bins cool.

Nalut is a Berber town, perched on the lip of the crumbling escarpment like Cubist flotsam on the shore of a gravel ocean. Some 30 kilometres south of the town the first dunes appeared – mounds of rilled orange sand. As we rolled ever southwards, wraiths of sand began to drift across the road, like fingers testing the hot tarmac surface. They were fed by peach-coloured dunes which had munched up the last of the scrub.

The camels, when they appeared, seemed entirely appropriate. We stopped, spellbound, as about a hundred and fifty of the strange, complaining creatures streamed past, the slanting sun flaring silvery auras off their shaggy coats.

Their Taureg herdsman shook my hand, then touched his chest.

'Salaam alaikum.'

'Wa-alaikum salaam,' I replied.

His face was the colour of the dunes and his eyes seemed to bore through mine to some ancient part of my brain. I sat, watching his departing back, not a little shaken by the contact.

Instead of the anticipated dunes, though, we rolled out onto a great gravel desert, varnished to an eerie sheen by wind-blown sand and stretching as far as the eye could see. To the west lay Tunisia and Algeria, to the east the trackless nothing of the Tarabulus region.

We are so accustomed to human ownership and use of the planet's surface that hundreds upon hundreds of kilometres of virtually untouched emptiness evoke a strange lightness of heart. Like the Antarctic, the Sahara is one of the greatest wilderness areas on Earth. Its sheer hostility has preserved it for the very few who know and respect its ways.

Beyond a tatty little oasis named Darj, darkness wrapped us in introspection. We'd been travelling almost dead straight for hundreds of kilometres.

'There are huge dunes here,' Sherif commented. But all we

saw along the tunnel of headlights were the occasional moth and a fleet-footed white jeboa, reminding us that even out here life still maintained a tenuous hold.

After the desolation of the desert the guesthouse in Ghadamès, Villa Otman Hashaishe, was a delight. A small restaurant nearby provided a fine meal, the shower was hot and the beds comfortable. I got a sense of what an oasis must feel like to a traveller after weeks on caravan.

Ghadamès is described as the Jewel of the Desert and its heart is an artesian well which has provided water for thousands of years. The old part of the town – being restored by Unesco – is an ingenious system of cool tunnels through a warren of multiple-storeyed houses and mosques.

Our Ghadamis guide, Mohamed Ali Kredn, was born in the old quarter and led us round unerringly. Without him we'd have been hopelessly lost inside a minute.

We left Ghadamès before sunrise under a full orange moon. The lights of the town soon disappeared and the dunes, rising and falling in the moonlight, looked like mid-ocean rollers. Dawn arrived quite suddenly in the moisture-leeched air. A glow on the horizon, then a ripple of bright gold which jiggled and formed into an enormous ball of fire dead ahead. The moon, like a pale lady afraid of the heat, slipped below the opposite horizon and was gone.

The day offered a sight both bleak and thrilling. The dunes had receded and we were crossing the Al Hamadah, a gravel desert so featureless that a molehill would have been an object worth studying. In every direction was a sort of biological and zoological nihilism, with the horizon as a near-straight line marked only by a colour change between earth and sky.

The arrow-straight road, silvered by the rising sun, seemed to leak sky into a widening flow around us and swallowed it back up behind our humming vehicle. The day's goal: 1,100 kilo-

Into the sands of silence

metres of absolute nothing. The boredom which this kind of road induces is an almost tangible thing. Your mind goes numb before the great emptiness, recoils upon itself and very soon begins to work in dream mode.

At an oasis named Ash Shwayrif the road veered south towards Sabha, the largest town in southern Libya. Brave bushes – strung like green pearls along dry wadies – were a welcome relief from the smashed-rock landscape. Future colonists destined for Mars could spend time in these wastelands getting themselves acclimatised.

I was so numbed I only focused on the trucks when one sped past us loaded with onions and smelling delicious. Others followed carrying wheat, maize and watermelons, all heading out of the wasteland towards the coast. Then came the fields of Sabha and the out-of-place zik-zik of water sprinklers. The town was not the dusty desert place I'd expected: the streets were lined with trees and hedges, and prosperity was obvious. The key to this mystery is artesian wells, dug thousands of years ago, and still producing the liquid of life in abundance. We pulled into a smart restaurant for hamburgers, kebabs and fruit nectar. I hoped we weren't eating camel.

From Sabha the road snaked along a wide valley with a harsh, high escarpment to the south and the edge of the massive Azzallaf Erg (sand sea) to the north. But, for more than 200 kilometres, the valley floor was covered with gardens and dotted with water towers, pines, palm and casuarinas. It didn't fit my Sahara stereotype.

We turned into Africa Camp at Ubari some thirteen hours after leaving Ghadamès. It's an attractive little tented camp with a restaurant at the foot of pink dunes, which stretched from east to west as far as we could see.

As night fell I trekked over a dune at least a hundred metres high and settled on the powder-soft sand to await the moon. A large owl glided past. The silence was profound, and if I had any travel tension it soon leaked away into the sand. Perhaps I sat there an hour, maybe two. Forty days and forty nights would have been just fine.

From Ubari we headed west along the foot of *mesak* (mesa) mountains. The dunes receded and we were once again on a sand sea. As we neared Al Awaynat, the desert turned from lemon to slate. Dunes occasionally appeared on the horizon, then drifted out of sight. In all that expanse there was nothing but pebbles, gravel and sand. When your eyes beg for variation, it's surprising how many colours you can pick out of this stark geography.

We swapped the Kia for a Toyota Land Cruiser in Al Awaynat and headed dead south, following tracks towards the Akakus Mountains. The sun was low, and when the foothills appeared they seemed covered in forest. But it was an illusion – out there was nothing but black rock and yellow sand.

Without road signs, obvious features or even tracks, the only way I can explain how we found the Taureg camel men is to say that our Toyota driver was also Targui. The desert men were camped in a sandy depression among crazy-shaped boulders. Their camels were hob-tied but looked placid and – if a camel has the capacity – happy.

Raia Abdul Alrhman Embarak greeted us courteously, shaking hands and touching his breast with the proffered hand. He was a desert-wizened man with clear, smiling eyes above his ever-present nose veil known thereabouts as a *litham*. But under his cloak you sensed steel. A fire was going and we were soon eating sand-baked bread and delicious Taureg soup.

'Camels', said Raia between mouthfuls, 'are my life.' His people have dominated central Sahara for centuries as raiders and caravan riders and they are still the desert's greatest fighters – true knights of the great sand ergs. As a youngster, Raia had crossed from Ghat in Libya to Niger with a caravan of dates. These days he does camel business in the Akakus, carrying archaeologists and occasional tourists.

Next morning he brought the camels down on their bellies ready to ride. I swung my leg over the camel's back, nearly impaling my calf on the spiked saddle. Raia yelled something at me and Sherif translated: 'He says grab the hair on the camel's hump behind you or you'll break your teeth on the pommel when it gets up.'

I made a grab as the beast came up, backside first, then gave a mighty shove with its front knees. This threw me backwards as the camel came aloft. The saddle spikes, I figured, must be a way of separating the Targui from the tourists.

We set off for the deep crags and valleys of Akakus. After several hours of travelling we turned up a wide wadi which led directly into the mountains. High walls of sandstone closed in as we lumbered up a virtual sand highway. The surrounding crags had the appearance of hammered sea cliffs. Wind-driven sand and time had transformed cracks and fissures into yawning caves, arches and canyons.

We finally dismounted near an overhang and gaped in wonder. The wall was covered in rock art. Here were gangly giraffes, elephants, antelopes and stretched, skinny people. In some scenes men with bows and arrows with hunting dogs pursued walia ibex, in others women braided hair or clapped their hands. Further up the valley was a trance dance conducted by foot-stamping and Bushman-like half-human, half-animal therianthropes. On the floor were tiny arrow heads. Had they once carried poison? At another site was a huge, perfectly proportioned elephant, and at still another were more recent paintings of camels and even chariots of ancient desert people known as Garamantes.

Who had painted these pictures? And when? The questions hung in the silence of the desert unanswered. We do not yet know for sure. Their style, though, was recognisable. I'd seen it in many caves and overhangs at the southern end of the continent.

It would be incorrect to say that the trip back to the coast was an anticlimax. We went by a different route and saw other places: the desolate oasis of Brak, the flower-filled Berber town of Ghariyan, the ancient Roman city of Sabratah ...

By degrees the healing silences of the sand ergs were replaced by the bump and chatter of urban life. But back in the silent valleys of Akakus questions remained. Had the Bushmen once been masters of all Africa? Was it they who had crossed the land bridge into Europe? Are we all their descendants? The evidence would need more work to be conclusive. But it was compelling.

One thing is for sure, if some of the extraordinary rock art in those mountain galleries is San, they were certainly great travellers.

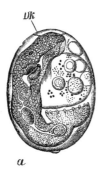

Naturally Nyalaland

There's not much call, these days, for the ancient fear of being eaten. That may explain why some people go out there and tempt fate with drooling lions or mean-tempered buffaloes. Just to remember how it once was before we crowded into cities and put our predators in zoos.

I obviously had all the wrong motivations: right then I was stomping round the Nyala Wilderness Area in the Kruger National Park because of my job. No hint of nostalgia for my naked ape ancestry. Which is why the strong bovine whiff had made me nervous – and remembering the buffalo story from the night before didn't help.

In the comfort of the camp round a crackling fire, old Daniel Maluleke – who's been shepherding hikers round Nyalaland for the past twenty years as deputy trail ranger – had spun a tale which should have sent us packing.

He was out with clients and a new senior trail ranger one day when a great black mountain with horns and intent thundered out of the bush. The buffalo snapped Daniel's gun and sent him flying. The clients bolted. He expected his colleague to shoot the creature, but the latter was nowhere to be seen.

Daniel dived into a cleft between two rocks and the buffalo loomed above him, trying to hook him out and do him in. A horn caught his leg and ripped it, but the beast wasn't getting satisfaction. Daniel desperately grabbed the buffalo by its nostrils as it peered down at him. That really made it mad. It rammed one of the rocks so hard, the thing split in half. Dazed by this excess, it then wandered off, its head no doubt spinning, and Daniel lay there until the other ranger and the hikers cautiously appeared to see if he was dead.

'Buffaloes', he concluded, 'are not to be trusted.'

Well, there we were in buffalo territory with a strong bovine pong all about and nothing but the sound of birds and my beating heart. Daniel and the other trail ranger, Rob Thompson, slung their rifles off their shoulders and kept walking while we all trotted nervously behind like a line of school kids playing a tiptoe game.

The smell receded, the danger passed and we all started talking, which you're not supposed to do on game trails. My colleague David Bristow – who was once a game ranger himself – frowned ferociously at us, but the rest of the party were Australians, so what can I say?

For most of the past decade northern Kruger suffered a terrible drought and most of the nyala had died or emigrated, together with most other species.

Then the rains had arrived in typical African overabundance and the bush, by the time we arrived, was thick, green and gorgeous. The trouble was that this made birds and beasts difficult to see.

An expert on birdsong, however, would have had a field day. Our morning had begun – way too early – with the sound of a rusty gate in the wind, which turned out to be a francolin. That woke a woodland kingfisher, which responded with a piercing 'trrp-trrrrrrr' that went on and on and on …

As the air warmed, a greenspotted dove counterpointed its plaintiff 'du du du-du-du-du' with the mournful refrain of a greenspotted wood pigeon: 'My mother is dead, dead, dead, dead.'

Then, as we hiked northwards towards the foot of the Maravula Hills, a black-headed oriel greeted us with bubbling oboe notes and competing red-chested cuckoos sang their Piet-my-vrou praises in two-part harmony. Given the foliage, we didn't encounter a flying thing.

'I can't see the damn birds,' I grumbled to the hiker ahead, but Bristow caught my eye so I lapsed into silence again.

That all changed as we threaded down Marivala Stream to a spectacular lunch-time perch above a waterfall: the secretive valley below was alive with birds. Daniel ticked off trees: 'Limpopo ironwood *(Olea macrocarpa)*, paperbark *(Melaleuca styphelioides)*, large-leafed rock fig *(Ficus soldarella)*, mountain mahogany *(Entandrophragma caudatum)*, brackthorn *(Acacia robusta)*, red bushwillow *(Combretum apiculatum)* …'

On the hill beside the valley was a *forest* of baobabs (that's not hyperbole) and above all this floated a pair of black eagles. It was the sort of scene that put drooling lions and mean-tempered buffaloes into perspective. We all made appreciative noises, stretched out limbs over the hot rocks, then chatted and chewed our way through lunch.

Seven hours after our dawn start we were back in camp, whacked. Some showered and slept, some just slept until the supper-time drum summoned us to the fire. The meal was oxtail stew, *pap*, vegetables and salads – a fine, robust way to end the day.

Next morning we were up with the rusty gate and the incessant kingfisher to explore the Luvuvhu River, a tributary of the Limpopo. Until recently it was a sedate stream, but the floods had reamed it out and changed its course in some places. The wild waters had even had the temerity to uproot some ancient sycamore figs and even baobabs. In the mud along its banks were the knuckle marks of baboons and the slide trails of crocodiles. Not a good place to swim.

On the way to the turnaround point we encountered the tracks of giraffes, civets, gannets and hares as well as a fat and beautifully marked puff adder, which lay beside the path with a just-you-dare look. The tangy aroma of wild mint nearby gave the sighting a delicious ambiguity.

We sat beside the river for a while, nibbling on ProVita crackers and cheese, while William Mabasa, who is Kruger Park's Media Relations Manager, instructed us on how to make a shepherd's flute from a dry reed. 'We made them when I was a youngster,' he told us. 'These days kids just listen to the radio.'

While threading our way back downstream, an oxpecker suddenly gave a loud 'zzzzzzist.' Believe me, it's not a sound you want to hear while walking in a Big Five game reserve. Both rangers immediately held their rifles at the ready and instinctively lowered their profile. 'What's up?' several people wanted to know, but Bristow gave them a warning look that cut short that noisy line of enquiry.

There are no oxen in the Kruger: oxpeckers mean big game's around and their scolding was an alarm signal warning their hosts of our approach. The air was again rich with the weedy, bovine smell of buffalo.

We advanced in our tippy-toe formation until the bush ahead gave a heart-lurching crash and a muddy, mean-eyed buffalo burst out only metres away. In that moment of adrenalin clarity I noticed there wasn't a convenient rock cleft in sight, only the crocodile-riddled river. But the beast thundered in the opposite direction and we all began babbling in relief.

Bristow sighed and asked if he could borrow my notebook. A few minutes later he handed it back to me with a neatly penned note headed *A Meditation on Reasons Not to Talk When Walking in the Bush*.

1. Human speech, more than any other noise, alerts and disturbs wild animals. So you see less.
2. The rangers need to have their senses clear to read the bush and be aware of danger. Chatting interferes with this.
3. Chatter brings to attention 'things' but blocks off your ability to 'walk like a hunter' – to assimilate the signs and sense of the bush.
4. Talking brings with it the rush and clutter of urban human life. Wilderness is about silence.
5. It irritates other trailists.

He was right, of course, though in humans speech is often faster than thought. I very nearly said 'thanks'.

But then I glanced at the stoically silent rangers, remembered the drooling lions – probably by then with their golden eyes fixed on our party – and held my peace. To a hungry lion the racket just then was probably the nearest thing to a flashing McDonald's neon – and I remembered the shortage of nyala.

Absolute Orange

'Here,' said Peter Bassett, stabbing his finger at a place on the map tangled in crazy contour lines but showing no roads or even tracks. 'The source of the Orange. Right on the edge of the Drakensberg escarpment. I reckon we could get you to within 15 kilometres of it by vehicle. You could hike the rest.'

As I sat in my office poring over a 1:250,000 map, the idea of following the Orange from source to mouth had seemed perfectly sane. You just flipped a few pages of the map book and there you were, on the coast at Alexander Bay.

Several months later, high in the Maluti Mountains and bounding over a bridle path which passed for a road in an old Series One Land Rover, the plan seemed a whole lot crazier. The emerald-green shoulders of the Malutis towered above us and far below, slithering in an immense valley of seemingly endless oxbows, snaked the Senqu – the start of a river which flows clear across South Africa and enters the Atlantic 2,250 kilometres away. It looked a heck of a lot more daunting than that thin blue line on a map.

In Lesotho it's more appropriate to talk about riverscape than landscape; the whole country is a westward-tilted watershed formed by uncountable arteries and veins of river and stream. It has the highest lowest point of any country in the world.

To get to the source of the Orange from South Africa requires some really strenuous hiking up the Drakensberg through either Mnweni or Rockeries Pass west of Bergville. But logistics dictated that we go up from the Lesotho side – where the Berg is known as the Malutis – approaching the escarpment from the river-braided south. That's big, high, wild country.

I spent the first night at Malealea Lodge, the best starting point for any adventure in Lesotho, then drove eastwards to Molumong Guesthouse – just south-west of Mokhotlong. Both places started life as trading stations and still retain the atmosphere of outposts of civilisation in wildest Africa.

We left the comfort of Molumong at dawn in two Series One Landies and my vehicle, bumping our way through the town of Mokhotlong and along an increasingly alarming track heading north.

'Stay east of the Senqu,' old Gilbert Tsekoa, who owns the store at Molumong, had told us. 'Maybe you'll get there that way.'

It is fair to say that many 4x4 drivers probably couldn't have got there that way, but Peter Bassett and his Conical Hat Expeditions team – Louis Powell and Neville de Klerk – are on a scale somewhere between brilliant and suicidal, qualities which appear necessary behind the wheel in Lesotho. My vehicle was being driven by Glenn Jones of Malealea Lodge, a man born of these mountains and bearing the brunt of Louis's evaluation of the double cab as 'Tupperware'.

It took most of that day to negotiate just 60 kilometres, and on a whim I'd chosen to ride with Louis. There were times – let me be honest – when I braced myself between seat and dashboard, closed my eyes and waited for the lurch into a precipice. A line from the 23rd Psalm about valleys and shadows of death kept recurring in my brain like a stuck record. But these hardcore Landie jockeys don't often make mistakes: they're a rare breed. My Colt just stayed in their tracks and kept coming: much, I suspect, to their amazement.

After fording the Senqu five times – with water sloshing into the cab twice – we finally arrived at a path up a hill even the redoubtable Louis had to admit was the end of the road. By the map we were, indeed, some 15 kilometres from the source – as the crow flies. But by bridle path, given the endless river rambles, the trail was probably twice that. As three of us headed north with packs on our backs, Conical Hat set up a base camp of muddy vehicles and bright blue tents and settled down to await our return: backpacking wasn't their line of business.

Following the river turned out to be a mistake. For reasons which might excite geographers, the Senqu has seen fit to oxbow off the escarpment, changing course every kilometre or so. It looked like a silver snake. It even hissed like a snake. I hoped this wasn't an omen.

We finally realised it was easier to go over the shoulders, high as they were, and there found a very creditable path. We later unearthed the reason: it's the Dagga Trail from the cannabis fields of Lesotho to the dope-hungry consumers of KwaZulu-Natal.

At nightfall we begged a place for our tent right next to a remote hut – it was the only level ground in bizarrely steep country. At dawn we temporarily abandoned the tent and sleeping bags, striking up the river with lightened packs.

After five hours of incredible scenery and really tough hiking, the path went dead on us in a blind canyon. We stared at the impossible cliffs, dumped our rucksacks and sat down. Our quest seemed thwarted. We chewed despondently on some Snacker bars, humped the packs and prepared to head back down.

Glenn Jones, however, is a mountain man who does not like defeat. 'Let's see what we can spot from that peak,' he suggested. 'Just one last shot.' It wasn't an easy climb – especially at an altitude of more than 3,000 metres – but it seemed a half-good idea.

As we crested the peak the view was spectacular. Below us a small stream emerged from a spongy swamp: the cradle of the Orange. All round, the rippling green mountains seemed to caress its banks. Beyond the stream clouds hurtled vertically into the sky, torn to shreds by high winds slamming them into the jagged rock needles above Rockeries Pass and against the escarpment wall.

'Phew,' said Glenn, sitting down suddenly. 'That's one heck of a sight!'

As with many high mountain goals, however, time and temperature left little time for awe or philosophical reflection. Snow had been forecast and we'd have to walk hard to reach the safety of base camp by nightfall. We turned our backs and headed downhill: the journey to the mouth had begun.

Base camp slid into view as the mountains turned from purple to black. A meal and a nip of whisky gave us insufficient fortification for what was to follow: the night was the worst I can remember. The temperature plummeted below the level my sleeping bag claimed it could cope with, the camping mattress was too thin to ease the pain of two days of hard hiking in thin atmosphere, and as sleep eluded me the world beyond the gossamer skin of the tent crystallised into hoarfrost.

At dawn we staggered around numbly until the sun began its healing work. A pair of socks, left out to dry, had frozen solid and had to be prised off the bumper of the Landie before being tossed in the back with a resounding clunk. We rolled up the steaming, soggy tents and began the slow, grinding ride back to Molumong Guesthouse where the thought of soft, warm beds, a place to dry the tents and, more importantly, a hot shower lured us on. They were there, and we luxuriated.

Next morning we headed south towards the village of Sehonghong. But just before the village the Senqu River presented us with some irresistible rapids. We pumped up the croc which Ark Inflatables had bequeathed us for the trip and shot down them in a welter of spray.

The sky wheeled overhead. As the craft was drawn into the main channel, I had the sensation of sliding down the vast tilted face of a country. Under me, in dancing whirls of sand, was the immense body of the continent itself, flowing, like the river, grain by grain, mountain by mountain, down to the sea. Whatever vehicle or vessel we chose, the Orange was clearly going to be a wild ride.

Beyond Sehonghong the road deteriorated into a 4x4 purist's delight. It literally tipped our vehicles into the Sehonghong Pass, which turned out to be somewhere between a cliff and a river bed. Idling speed in low range had little effect in stopping the Colt from sliding down sheets of rock towards terrifying hairpin bends.

Strangely, we survived, forded the river, then hacked our way up the other side of the canyon. If my eyes hadn't been glued to the road and my mind focused on raw survival, the scenery could have been competition-photo stuff.

The road skirted Mashai Lodge, a good place to overnight. But we decided to camp, and dropped to the confluence of the Senqu and Matebeng rivers where we pulled under some ancient poplar trees. By eight o'clock we were in bed, leaving the night to busy crickets, a skinny dog which decided to defend our camp, and, the gurgle of the unceasing river. The

American philosopher Henry David Thoreau once said: 'Who hears the rippling of rivers will not utterly despair of anything.' He knew what he was talking about.

Next morning we began climbing the Matebeng Pass to a neck which seemed to explode views upon us both east and west. I could almost hear the trumpets of Beethoven's Fifth in the icy dawn breeze. This crazy road, which has not yet succumbed to the blessings of modernisation, offers some of the finest 4x4 adventuring in Southern Africa. And the scenery! ...

Near the village of Sehlabathebe we took our leave of the Conical Hatters (in their terms the difficult stuff was over). Peter walked round the Colt, kicked a tyre, and conceded: 'It sure earned its spurs up there.' Then he hopped in his overgrown shoebox and roared off. Our team was now down to three, the other two being Di Jones of Malealea Lodge and Neil Bennun, a journalist friend from London.

Light was falling but we pressed on in search of a secret Senqu gorge which Di knew about near the village of Seforong. A sign to Hareeng School was the only clue and, under Di's instructions, we dived off the road down a steep, rocky track (4x4s only) and ended on a grassy sward beside a stream under massive black-and-yellow sandstone cliffs.

The campsite could hardly be more spectacular. As darkness stole in, the Milky Way turned on like city streetlights – anchored at each end by massive rock walls. Next morning, though, became a hectic scramble to beat an early storm which dumped on us as we threw the last of the gear in the Colt, dived in and slammed the doors. Rain hammered on the bonnet as we low-ranged our way out of the now-slippery gorge.

We'd missed breakfast in the rush and were rather hungry. On a whim we turned in at the Mphaki Farmers' Training Centre which appeared over a rise, and were treated to maize porridge, eggs sunny-side up, fat pork sausages and tea for a few rands. The rest of the guests were, indeed, farmers.

From Mphaki we motored along the Senqu, past Mount Moorosi and hugged the Senqu to Quthing through seemingly unending, jaw-dropping valleys, then turned off towards the border post at Tele Bridge. Soon after our passports were stamped, the countryside flattened out and by the time we reached Aliwal North the mountains were mere memory. We were already missing Lesotho.

Aliwal got its name by one of those curious colonial quirks and a desire to please a high official. Cape Governor Sir Harry Smith had to his credit a battle in India at Aliwal: one of his accolades was 'Hero of Aliwal'. South Africa already had Harrismith and Smithfield, so the town became Aliwal North. Why north? Well the town of Mossel Bay, at the time, was known as Aliwal South.

There's a curious cosmic but little-known fact about Aliwal which turned up in the fascinating local museum: the town remains perfectly still while the sun and stars revolve round it.

It happened like this: in 1872 a dominee of the NG Kerk revealed to his shocked congregation that the earth rotated round the sun. Maybe he'd sneaked a look at a book on Galileo. A Mr J W Sauer got up in the middle of the service, objecting to this rubbish, and walked out – followed by most of the congregation.

An urgent meeting of the *Kerkraad* (church council) produced a resolution they figured would win back the congregation. Whether they believed it is not recorded: 'As from today the earth round Aliwal North no longer rotates.' The order has never been rescinded and, apparently, is still valid. So, whereas the rest of the world busily revolves, Aliwal is standing dead still.

A better-known feature of the town is its thermal springs. The baths built round them consist of a warm pool and several cooler outdoor ones plus an extremely long water slide. The place is in need of a makeover but the water is still warm and healthy. An old advertising poster in the museum describes the baths rather unsettlingly as 'Radioactive Hot Springs: Most Efficient in Curing Rheumatism, Gout, Paralytic Afflictions, Sprains and Skin Diseases'.

Accommodation for self-caterers at Oppie Bron next door offered a cosy chalet at backpacker rates. It was a good opportunity to do some laundry, but the clouds clogged the sun and we had to bundle soggy kit into bin bags, hoping for better weather further on.

We drove over the Orange River on an old iron bridge from Aliwal into the Free State, then turned towards Bethulie. Before reaching that town, however, we came across a game reserve I'd never heard of named Tussen die Riviere, situated between the Caledon and Orange rivers. Its deserted roads suggested that quite a few others hadn't heard about it either.

It's a beautiful place of koppies and grassland. The ranger in the camp office looked a bit startled to see us, and even more startled when we asked if we could use the park's tumble dryer. But he offered it to us with good grace and we lunched on the lawns to the hum of brief civilisation.

On the way out of the reserve we spotted a cave which just had to contain rock art. After a stiff climb we discovered that it did: a remarkable San trance-dance scene. May it remain unknown for as long as possible – given how rock art is generally being defaced.

The Gariep Dam – the first barrier to be encountered by the Orange – began to make itself felt as we drove past tongues of silver water lapping the dry hills between Bethulie and Donkerpoort. Its 900-metre-long wall was completed in 1972 and, when it's full, the water covers an area of nearly 400 square kilometres.

Nothing alters a river as totally as a dam. After a week of travelling along – and on – the Orange it had become, for us, a living thing. We'd delighted in its garments; we knew its soft voice and were willing captives of its restless spirit. As the waters of the Gariep widened and the river slowed, I felt a strange desperation. I think it was the river's.

With directions in hand, we turned off the road near the wall and down along the river to De Oude Pomp and the hospitality of Ivan and Marlize Sinclair, who also run Big Sky Adventures.

After a hearty Karoo meal cooked by Marlize and a cosy night in a tented chalet overlooking the river, we headed for the towering dam wall and Ivan's put-in point for some genteel kayaking back to De Oude Pomp. Some questioning along the way established that the giant wall was a honeycomb structure worthy of adventure.

Dam official Neville Swarts offered to show us round. We stopped on the wall to wait for him and were surrounded by cats. Most seemed to emerge from a single dustbin, others from gutters and out of bushes. They padded round us for a while, then disappeared – river spirits, obviously.

Neville's first bit of rather disquieting information was that the dam was 98 per cent full and, as a result, the wall had moved seven millimetres downstream. It wasn't much, but standing inside the gargantuan thing on bedrock 88 metres below the top (with one foot in the Cape and the other in the Free State) it wasn't too comforting. We popped out of a door near the base of the wall and he opened a massive sluice to give us a taste of raw water power. The deafening roar echoed off the valley walls.

We crossed the busy N1 highway at Donkerpoort, ignoring the brief, rumbling reminder of cities and traffic, and hit a duck. I don't want to get sentimental about this, but the plump, white farmyard quacker was sitting in a rain puddle in the middle of the road and I thought it was a plastic packet. My first recognition of it was startled yellow eyes metres from the bull-bar, my last was a cloud of feathers in my rear-view mirror. We were to discuss that duck often. I felt very bad about it all the way to the mouth of the Orange.

Post-duck, we found ourselves in Philippolis – one of the most charming villages we were to encounter. Its buildings and neat gardens seem little changed since its mission days in the mid-nineteenth century. The town was established by Dr John Philip of the London Missionary Society. For a while it was the home of a section of the Griqua people under Adam Kok before the group embarked on an incredible trek to found Kokstad just west of Port Shepstone. That's worthy of a story all to itself.

The town was also the chosen site for a spinning school for Boer women set up in 1905 by the tough-minded Emily Hobhouse, fresh from exposing the outrages of British concentration camps during the Anglo-Boer War.

This historical information – plus some dangerous-tasting witblits – came courtesy of the Trans Gariep Museum, which is a model of what small-town museums should be.

Round the corner is the house in which the writer Laurens van der Post grew up. It's now owned by ex-Gautenger Mark Ingle, who's gone to ground there to help pen the writer's biography. No finer place for inspiration than his subject's house.

We reached the Vanderkloof Dam north-west of Philippolis in the slanting light of late afternoon and it was eerie. It's the second great barrier across the Orange and the wall was certainly impressive, but it had the unsettling air of a deserted stage set: koppie-studded backdrop, tourist facilities, fine roads but not a soul in sight. Do robots run it? Which planet were we on?

It seemed a good place to turn our thoughts to the *volkstaat* of Orania. We hit the R369 and gunned the Colt into the setting sun. Ahead was the only *boerestaat* on earth. We hoped we'd be welcome.

The *volkstaat* of Orania. It conjured up images of khaki-clad *boere* with Voortrekker beards, tannies in *kappies* making sticky-sweet koeksisters. Were we heading for the heart of some future ethnic republic? More likely, for a laager of grumpy, no-hoper

Afrikaners in the middle of nowhere resisting the new South Africa.

'Do you think I'll get into trouble there?' asked Londoner Neil Bennun nervously. 'I don't speak any Afrikaans.'

We seemed to drive across dead-flat tableland for ages; it was like being marooned in mid-ocean. When Orania appeared it wasn't in a 'there it is!' sort of way. It shimmered into view, surreal and floating, in an afternoon mirage.

A large man in khaki with a winner of a beard – fulfilling all our expectations – directed us to Die Herberg guesthouse. Signs along the way informed us of the value of one's own labours and the sound sense of settling in Orania.

Janneke Steyn, who opened the door, wasn't wearing a *kappie*. In fact she'd do just fine on a modelling ramp. '*Kom binne*,' she invited, leading us to rooms decorated in the finest Biggie Best. On the walls were framed – and seemingly original – cartoons *lampooning* the creation of the *volkstaat* and a collection of Pierneef etchings which must have been priceless. We sipped whisky on leather loungers and partook of a fine dinner (which certainly wasn't a braai), then turned in a little shell-shocked from collapsing stereotypes. On first meeting, Orania seemed to be very … sophisticated.

Like many towns along the Orange, Orania didn't use the river for much more than a supply of water. It had been built for workers constructing the Vanderkloof Dam, deserted on the dam's completion, then later bought by a *volkstaat* consortium. Requirements for entry are that you have to be a genuine Afrikaner.

'Brown Afrikaners?' I asked Kokkie de Kok of the Orania Kultuur Historiese Museum.

'Why not?' he countered.

His museum has an impressive collection of old guns, Anglo-Boer War memorabilia and the suit former State President and apartheid hero Hendrik Verwoerd was wearing when he was stabbed to death. Little red arrows show the places where the knife went through.

Orania also has a high-tech dairy – automated to reduce labour – a shady campsite, hectares of pecan nut and olive trees, a house built of straw (with no wolf in sight), a modern school with state-of-the-art computers, a jeweller who's as pretty as the delicate pieces she crafts, and Tienie Hout, a 69-year-old former bank employee who now carves huge tables and chairs out of impossibly hard, gnarled wood. Everyone was keen to show us round and offer us good, strong rooibos tea. I suspect the place has the highest percentage of tertiary-trained citizens in the country. Let me be frank, Orania was a complete eye-opener.

Fifty kilometres up the road – at Oranjerivier Station – we came upon train control officer Charl Mans, whose switching box was filled with railway memorabilia and an array of gleaming levers which would make a train enthusiast weep with envy. As a big diesel rumbled into the station, he pulled the levers to switch points with a white cloth in his hand.

'Sweat makes the handles rust,' he explained.

We hit Hopetown, further up the road, at one o'clock and, as we ran for a café, its doors slammed shut. So did all other doors of succour, so we sat in the park, ate dry biscuits and cheese, and reflected that the discovery of a 21-carat diamond there in 1866 – then one of 83 carats two years later – hadn't brought the town much lasting development. Maybe it was a judgement based on hunger.

Just outside Hopetown, though, is a surprise in the midst of dry, forbidding terrain. The Toll House is in fact two houses in a garden of lawns and flowers beside a historic, cast-iron wagon bridge over the Orange. You can stay there, watch the river slip by and drink in the head-spinning silence.

From Hopetown we followed a rather good dirt road south of the river to Douglas, an agriculturally rich town near the confluence of the Vaal and Orange rivers. It has the distinction of being South Africa's dead centre – geographically speaking. There's an excellent campsite just out of town on the banks of the Vaal, which offered a spectacular sunset.

Just before dawn I headed for a place named Bucklands, where the two rivers meet. The air was icy. I accelerated up behind a bakkie driven by a large, white farmer and, from the steam on his windows, the heater was doing sterling service in the otherwise empty cab. In the open back, however, was a dark-skinned woman in a tatty, thin cardigan who looked as though she was freezing to death. In the platteland old habits die hard, it seems.

From Douglas to Prieska is a long, flat road – which made me think about the horse latitudes. They're an area of the planet characterised by baffling winds, calms and high barometric pressure. On the old sailing ships horses used to die at these latitudes for no apparent reason. At 30 degrees south, Prieska is right in them, which may have explained the absence of horses. The town takes its name from *Prieskab*, which means 'place of the lost goat' in Koranna. There's no shortage of those.

There was also no shortage of tea and toasted sandwiches at Annie's Coffee Shop, where we got directions to the gem-polishing factory where the town's biggest export, tiger's-eye, is worked over. Inside, women were cutting the stone with dangerous-looking saws but the baas, they said, had gone to town. His office door had a sign on it which read 'Beware of snakes'.

It was in Prieska that the botanist William Burchell first came upon the Orange in 1811 and was enraptured by it. 'The first view to which I happened to turn myself, in looking upstream,' he wrote, 'realised those ideas of elegant and classic scenery which are created in the minds of poets, those alluring fancies of a fairytale or the fascinating imagery of a romance.'

Some ten years later that great slaughterer of African wildlife, Captain Cornwallis Harris, was to plagiarise Burchell's words to describe the same scene – which proved he was not only a butcher but also a thief.

From Prieska our road took us through Koegas, then to within a few kilometres of Marydale, where we turned north again along a wonderfully scenic road to the Boegoeberg Dam. The dam is a smallish barrier with a long wall and was spilling at the time. You can camp beside it and there are a number of chalets for hire. The young man at the campsite gate looked at us as though we'd just pulled in from Venus. I guess we were out of season.

The road below the dam followed a canal for many kilometres before depositing us on the N10 to Upington. We picked up some pretty good, inexpensive wine at the Oranjerivier Wine Cellars just north of Grootdrink (it seemed appropriate to do so, given the name), then made for the marvellously appointed Die Eiland campsite along the Orange in Upington.

Overawed by the size of the town and the bustle of traffic, we decided to spoil ourselves at Le Must Restaurant and were very glad we did. The food was a treat after our camp cooking and, while it wouldn't be correct to say that we became drunk and disorderly, we did wake up next morning with heavy heads.

A few aspirins and a coffee later we were in the local graveyard using nothing but intuition to locate the grave of the legendary nineteenth-century bandit Scotty Smith. We found it, by chance, then set off in triumph to survey his domain along the Orange. First stop was Kanoneiland, where the cannon which drove river pirates from the island is still on display.

At Keimoes it seemed a shame to stick to the boring N14, so we turned south to Neilershof and took the scenic, winding dirt road to Kakamas to view some old water wheels which are still at work, dipping water out of the canals and onto fields.

At Augrabies Falls, just up the road, the Orange was growling into the granite gorge, throwing up an impressive rainbow. Instead of sleeping at the park, though, we dropped in on the Kalahari Adventure Centre just upriver and were offered a room and a braai.

Early next morning we hit the Orange with a gaggle of crocs. Despite thorough schooling from river guide Johan du Preez, the first rapid – which thumped us down a hurtling vee of water into a snarling standing wave – had us yelling in fright and pumping adrenalin. If I tell you the next four rapids were named Rodeo, Klipspringer, Blind Faith and Beginning of the End you'll understand that the following few hours had us appreciating the raw power of our river.

We finally portaged the inflatables to a quiet side stream and paddled across it only metres from the start of a 150-metre boiling rapid which ended in the Augrabies Falls. It's not a stream you want to make a mistake while crossing!

After a quick shower we set off for Pofadder. On good advice we headed for the Pofadder Hotel where, we were told, we could get permits and a map to do the Namakwa 4x4 Trail from Pella to Vioolsdrift. The hotel was closed, being Sunday, so we rang the bell. After a long pause a sleepy voice in the intercom said, 'Ja?'

'Is the hotel open?' I enquired.

'Nee!'

'We need a map for the Namakwa Trail ...'

The speaker clicked off and I stared at the door, unsure what to do next. It suddenly opened a crack and sneezed. The nasal explosion was the work of a woman who confessed to having 'flu. Despite her resident virus, she was happy to give us all we required plus another hearty sneeze.

Pella is a neat village with a peaceful mission station at its heart. We hoped to overnight there but couldn't find anywhere to sleep, so we headed for Klein Pella, which turned out to be a lot bigger than Pella itself. It's an 18,000-hectare date-and-grape estate on the banks of the Orange with a guesthouse billed as 'the best overnight accommodation in Bushmanland'. I couldn't argue with that: we were given air-conditioned huts, naartjies off the trees and the largest breakfast of the trip.

Klein Pella nestles among harsh, desert-red mountains and has more date palms than anywhere else south of the Sahara. From there, after some confusion about just where it began, we linked up with the trail. There's a new section from Pella to Klein Pella which we were told about later and, given the scenery, we wished we'd known about it earlier. Silver-white bushman grass shimmered in the sun, softening the stark mountains, which seemed to pile higher and higher as we headed west. After perhaps a thousand kilometres of wandering the flat Karoo plains, the river was once again plunging into berg country. A funny thing for a river to do ... and what wild mountains!

The trail was not too punishing and relatively well signposted, leading through fossil rivers – old feeders into the Orange – and down valleys decorated with quiver *(Aloe dichotoma)* and camelthorn *(Acacia erioloba)* trees. It eventually descended to the banks of the Orange where we promptly got stuck in deep, powder-like sand. At two bars the tyres just dug in. There was no shortage of rocks, and after some dusty digging and a reduction of tyre pressure to one bar, we were soon on our way again.

There are many stories about this wild part of the river, the most persistent being about a huge snake with a diamond on its forehead which patrols the lower reaches. We didn't see the snake. But the hissing river soon led us to a spot marked on the map as 'wonderboom'. It turned out to be an immense, ancient wild rock fig *(Ficus cordata)* which offered delicious shade within fifty metres of the river. Sunset turned the mountains from gold to red and we broke out a gas wok for a riverside stir-fry. That night I dreamed of diamonds.

As the dawn sun bathed the valley, the huge tree shimmered and glowed like a living jewel. We left it reluctantly, heading away from the Orange up a spectacular fossil river bed, then curving back to rejoin the brown waters near a village once named Gudaus – now corrupted to Goodhouse – where the Orange was first seen by a European (a farmer named Jacobus Jansz who was on a hunting expedition in 1760). In January the temperature there can get up to around a blood-boiling 50°C.

A little further down the trail is the odd little village of Henkries, from where you can either take a good dirt road to the N7, or stay on the trail and join the highway 20 kilometres from the Vioolsdrift border post.

After 260 kilometres of wild trailing, the highway and the border post were a bit of a comedown. But border paperwork was brief and friendly and we were soon in the Felix Unite river camp with its green lawns, hot showers and more than 200 Mohawk canoes with their bows all pointed down river.

The camp is run by the affable Carlos Peres, who joined us for a sunset whisky and beguiled us with stories from his 15-year acquaintance with the river. Felix takes around 3,000 people down this stretch of river and other operators probably send down that many again. That's not enough: it should be compulsory travel for everyone within 5,000 kilometres of the place.

After a leisurely post-breakfast start we hit the first rapid with some trepidation. We'd joined a party of older river trekkers – and loaded Mohawks aren't as manoeuvrable as inflatables. Our guides, Yango John and Richard Goodman, set up a good deal of playful tension before the rapid – but it served to keep us alert and, on the first day, nobody capsized.

The mountains hereabouts could keep a geologist happy for a lifetime. Without a scrap of plant matter you could see the very bones of the planet: contorted, pocked and even spiralled. As the sun sank, the mountains seemed to clothe themselves for a carnival: red, orange, purple and even bright fire-yellow.

We pulled onto a spit of sand below a hill looking remarkably like a gorilla and the guides were soon serving spatchcock chicken, rice and salad, with fruit salad and custard for afters. It's amazing what supplies you can carry in a Mohawk.

The second day was slow, lazy and hot, the river flowing sluggish and brown beneath our hulls. We lashed the canoes together and leapt into the cool water.

I drifted away from the cluster and floated down river on my back, becoming one with the river. The only sounds were occasional laughter from my fellow travellers and a gentle rumble. The sheer bliss lasted for a few minutes until an alarm bell went off.

Rumble?

Rapids!

I flailed back towards the canoes and hauled myself on board, scrabbling for a life jacket and paddle as the roar grew ever louder.

Sjambok Rapid is so named because the whole wide river is squeezed into a channel perhaps eight metres wide. The sluggish water becomes fast and angry within this rude constriction. The canoeists ahead of me, unable to cope with the sudden fury, slammed into a rock and capsized. I struggled to steer my sluggish Mohawk over roaring waves, round sucking holes and past the floating contents of the upturned boat.

Canoes, flotsam and occupants all ended up shaken but unharmed in a quiet eddy below the rapid – except for someone's wooden bracelet which my paddling buddy fished out about a kilometre downstream.

That night, around three o'clock, I awoke and stared up at the canopy of glittering stars. The Milky Way stretched from horizon to horizon and the dust clouds cloaking its swirling centre were clearly visible. Meteors zipped this way and that like fireflies. It was the same fine sky I'd seen back in Lesotho, but out here in the desert it seemed to have been digitally reworked to perfection.

The final day of canoe trailing took us along the edge of the Richtersveld National Park. Beneath towering peaks we shot the final rapid – a roaring thing with rocks in its teeth named

Stairway to Heaven. There's something about the names they give these rapids ...

Our trusty Colt had been delivered to the take-out point and I rather reluctantly swapped my paddles for a steering wheel. From there the road hugged the Orange. A flood had taken out the bridge over the Fish River at its confluence with the Orange and we had a few tense moments fording it on a hastily bulldozed ramp of sand.

We entered the diamond area known as Sperrgebiet (forbidden territory) after a rather long wait at the gate. Radioactive-looking cards were issued and we were handed a list of rules which made it quite clear we weren't allowed to do anything to deprive the Namibian Government and De Beers Mining of a single grain of sand, never mind a diamond. The scenery was reminiscent of those pictures beamed back by the Mars lander. Was there life out there? Was there anything out there?

Well, yes: we crested a hill to find that a vast surface of desert had been worked over by giant machines and dumped back in ugly piles. Having been schooled in the tradition that deserts are fragile places, the landscaping exercise came as quite a shock. What we'll do to procure a girl's best friend.

Around 100 million years ago the diamonds were formed in kimberlite pipes of molten larva boiling to the surface and then washed into the Orange from side streams. Today they can be found in large numbers in the desert south of Rosh Pinah and all along the coast, where they've been scattered, by among other things, a process of longshore drift.

Dunes – pink, cream and orange – began to appear as we headed further west. We first spotted them nestling in the lee of blackened cliffs, then spreading out to become mountains in their own right. The river began meandering between arching banks of sand, seeming to pause before its moment of epiphany.

We went through a checkpoint, handed in our radioactive cards and raced into Oranjemund, looking for the mouth. But, despite the town's name, the 'mund' was elsewhere. Seeing a sign to the yacht club, we chased down the road beneath a setting sun, passing the clubhouse and ending on a spit of sand piled high with river debris. The Colt skidded to a halt and we all baled out.

To the north was the Orange, calm and resigned to its fate. To the south were the roaring breakers of the Atlantic, noisily consuming the brown waters of our river. We stood, staring at the meeting point. Then Neil Bennun drew an arrow in the sand pointing east and wrote the word 'source'. Our quest had ended. But up there, in the wet cradle above Rockeries Pass in the high Drakensberg, the long journey of a small stream was, right then and forever, just beginning.

b

Into the forbidden desert

The Blue Diamond Lodge in Springbok was a museum to whimsy.

It contained, among many other things, an ostrich-foot ashtray, a spinning wheel, a wind-up gramophone with a sign on it 'Thank you for not smoking', some 3-D paintings of zebras and suricates framed by logs and held in place by leather thongs, a Hammond organ adorned with copper jugs and pixies, a now-silent stone fountain (with dolphins) on the carpet, a statue of Akhenaten with snake headgear and another of Nefertiti. Outside was a fine garden dotted with eager-looking gnomes.

My bedroom was a masterpiece in crocheted doilies, with a fur-edged dressing table and two large paintings of young women, one wearing nothing but a sultry look.

The finest piece was the bed: huge, fur-lined and surrounded by mirrors, strategically placed lights and what appeared to be a handbrake, the purpose of which I was unable to discern. A fan circled lazily overhead.

As a counterpoint to what lay ahead, it would be hard to beat.

On the patio could be heard the incoherent roar of a party beyond the point of no return. 'There will be headaches,' I reflected as I lay back in fur-lined luxury, having taken early retirement.

There were. Breakfast was a groaning affair and the blazing desert which awaited us clearly held no attraction. Sprawled across several tables were Johan, Hercules, Dawid, Marius, Chris, Gavin and Pieter – an unlikely cast, right then, for high adventure. Jaco and Neels, who lived in Springbok, looked in better shape and chivvied the guys into the 4x4s and up the road to the Namibian border. At Vioolsdrift we stocked up on beer – 11 cases – and headed down the Orange River to Rosh Pinah Mine on the northern edge of the Richtersveld. The many hangovers made for a rather sombre trip.

We stopped beside the Orange for sandwiches, meatballs, juice and, what the heck, a few more beers. The wind funnelling down the river seemed to come from a giant hairdryer. Our sandwiches curled in protest.

Planning for the trip had begun a few months earlier with a call from Johan Kellerman of Safari Centre in Cape Town. He'd discovered that an outfit named Coastways in Lüderitz had landed a concession to take limited numbers of 4x4s into Sperrgebiet 2, the vast, formerly forbidden diamond mining area north of the town. Should we give it a go?

Chris Carolin of Porter's Auto was contacted for advice. He's a 4x4 man and offered no less than three new Pajeros if he and some buddies could come along. Good publicity, he rationalised. When Johan and I rolled into Porter's early one morning, a motley crew was waiting. It was clearly going to be a *Boy's Own* adventure ...

As we headed up the long, sandy road towards Aus the boys started perking up. We'd installed two-way radios in Springbok and the nonsense began.

'Perone, Perone, Perone, this is Chris Carolin here. Do you read me? Over.'

'Loud and clear,' chuckled Gavin Perone.

'Do you know why ducks have webbed feet? Over.'

'No. Over.'

'So they don't sink into the sand. Over.'

Groans emanated from the vehicles.

'Perone, Perone. Do you know why ostriches stick their heads in the sand? Over.'

'No. You tell me. Over.'

'To hunt for ducks without webbed feet. Over.'

By the time we got to Aus the humour was diving faster than an un-webbed duck in a dune.

The little town on the lip of the Namib Desert was once the railhead for southern Namibia and a concentration camp where German Schutztruppe were interned during the First World War.

Aus has a reputation for being the coldest place in Namibia: when writer Lawrence Green camped there some fifty years ago many of his camels froze to death. For us, it was cooking. We bought a few Cokes, stared into the immaculate engines for a while – the way 4x4 people do – then tootled down the long, straight road to Lüderitz.

In 1487, when Bartolomeu Dias arrived in the bay where the port would eventually materialise, it was a profound disappointment. For weeks he'd been battered by high seas and merciless winds. The dunes to the east never seemed to end, and when he sailed into the bay and found not a drop of fresh water it was the last straw. He erected a stone padrão on a rocky promontory, named the coast *Areias de Inferno* (Sands of Hell) and fled.

The place has improved a good deal since then. Lüderitz is a really attractive town with colourful German-style houses, some comfortable B&Bs and a raucous pub down at the yacht club. We made our way to the Kratzplatz Guesthouse and there met Volker Jahnke of Coastways, the man with a permit to Bartolomeu Dias's hellsands.

Next morning the vans were on the road again, whistling back towards Aus for about thirty kilometres, then northwards into the forbidden territory. Soon, around us was nothing but sand.

'Perone, Perone, Perone. Do you think it would be an idea to float a company in Lüderitz building sandpits for children? Over.'

The Chris 'n Gavin Show was on the air again.

'Carolin, Carolin, Carolin. I read you. I think the market might be saturated … Over.'

Volker chuckled into his ample beard. 'Have you had this all the way from Cape Town?' he asked.

'No, only from Springbok.'

'Ja, bad enough … '

Travelling into the Namib-Naukluft Park with the big, German-speaking Suidwester was an adventure in itself.

'I need my wind,' he pronounced as we roared up a giant dune in his 4.5-litre Land Cruiser, its flattened tyres growling against the khaki-pink sand. 'Without wind there'd be no desert – and the wind erases my tyre tracks.'

The three Pajeros and a supply Land Cruiser driven by Jaco were in our tracks. As we crested the dune our Cruiser's front end dropped off a near-vertical wall, flailed about a bit, all four wheels spewing sand, then dived down the slipface in a wedge of sliding dune.

'You gotta watch those little ridges,' grinned Volker as we watched a Pajero belly onto the dune crest. 'They're caused when the wind changes direction and builds little cliffs on the crests. If you hit them at an angle you can flip. Nasty way to come down a dune.'

One particularly mean dune – the sort of thing it wouldn't have occurred to me to challenge with a 4x4 – was crested by all but tail-end Jaco. His loaded Toyota wined, spewing gouts of yellow sand and shuddered to a halt on the slope.

'Dump the beer!' someone shouted.

'No way!' came a chorus of voices.

After a few more failed attempts Volker sighed, jumped into his vehicle and plunged down the slope to lead out the heavily packed Toyota by an easier route.

The desert before us was an undulating tawny blanket under a Chagall-blue sky. At one point we crested a rise to find a vast spread of dunes rippling to the horizon, at another we were thrumming along mere metres from a towering slipface which blocked half the sky.

Cloaking nearly the entire coastline of Namibia to around a hundred kilometres inland, the Namib is a place of spectacular extremes and lip-cracking aridity. An area is considered to be a desert if the evaporation is twice the rainfall. In the Namib it's around 200 times higher.

Swakopmund holds two world records: the longest documented rainless period (15 years) and the greatest rainfall variable (1 – 148 mm). This desert – thought to be the oldest in the world – is also the only place where fog-generated moisture can exceed rainfall. Vogelvederberg, a granite dome 50 kilometres from the coast, receives nearly ten times as much fog precipitation (183 mm) as rainfall.

With no plants or water to hold the fine-grained sand, the Namib has become an ever-moving sculpture of the south-west wind. Grains jump under its incessant coaxing, forming curtains of moving sand which prowl the dune fields, grinding down anything in their path. They cut grooves, plume sand behind objects, build dunes to more than a hundred metres high, and march giant, moon-shaped barchans at high speed over the gravelly desert floor.

We slid to a halt beside the sandblasted remains of a Jeep. All that was left was a steering wheel and the engine block, rust free but polished to gleaming black. Nearby was a metal bedframe with a shiny spring some way off, half buried. On the remains of a corrugated-iron hut a crow had constructed its nest entirely of wire. It struck me later I should have checked to see if that's where the other springs had ended up.

'In ancient times winds round here blasted at around 300 kilometres an hour,' Volker informed me as the Cruiser growled up a huge dome of sand, its wide Goodrich tyres drumming like thunder. 'Everything was ground to bits. These days a southwest blow can still bury a vehicle overnight.

'The sands here are 90 per cent quartzite. Those red patches in the dunes are garnets, the black stuff is ilmenite, on the beaches you'll see agate – and, of course, there are diamonds. It's a crystal desert.'

For some time we travelled along what Volker described as dune streets, conveniently flat valleys between huge, longitudinal seif dunes. The high crests on either side had constantly repeating S-shapes, rising and falling to form a chain of summits. At a point Volker seemed to intuit rather than see, we turned towards the coast and hit the really rough stuff. As the vehicles clawed their way westwards the dunes piled up to impossible heights. It was extraordinary to see where 4x4s could get to.

'What happens if a vehicle breaks down?' I asked.

'We fix it or we tow it out. But in these dunes you can't tow, so if you can't fix it you have to come back, cut it up and haul the bits out. We're not allowed to leave anything in the desert. That's part of the concession rules.'

'Have you had to do that?'

'Yep … '

'What about tracks?'

'On the gravel beds the tracks will last for hundreds of years,' said my companion. 'They're an eyesore. You can even still see hundred-year-old game tracks there. I'd never drive in the beds. But on dunes the wind erases all traces of 4x4s within 24 hours. You'll see. We'll come back this way and I bet you won't see a track.'

Saddle Hill North appeared suddenly without preamble. Hidden amid tough coastal bush at the edge of a dune field, its weathered shacks seemed half buried and far from welcoming. The place had been a diamond mine, where hardy men had scratched alluvial gems from the nearby beach. Volker had turned the workshop into a living room with kitchen and circular fireplace, and some of the offices into rudimentary bedrooms. It proved to be far more comfortable than first appearances suggested.

There was, however, a strangeness about the place. I looked into one of the unrestored rooms. A ridge of sand rose half-way to the ceiling like a deep-sea swell frozen in the act of swallowing rusty office shelves and broken machinery.

On the sand-polished roof two crows with wire-bending beaks held council like raucous undertakers to men's long-dead dreams of glittering fortune. Wind whistled round broken planking and hissed over sand rills, blurring the ground and driving sea mist inland over the thirsty desert. It had been a long and extraordinary road to this remote corner of nowhere.

After a fine meal, some liquid lubricant and yet more gags, we turned in to the sound of Atlantic breakers spewing Orange River diamonds onto the beach. We hoped.

Next morning we trundled out under billowing clouds of fog to Saddle Hill proper, a mine almost wholly swallowed by the moving sand. It seemed to be inhabited by only a family of black-backed jackals, though we saw merely their tracks. Rusting machinery lay near a hard-surfaced road which, it turned out, was an old airstrip.

'Hey, Chris!' yelled Gavin. 'Check out this old steering wheel.

There's nothing else left. Must have been a Pajero … '

From the mine we headed northwards towards Spencer Bay through scenery which must rate as some of the most extraordinary in the world. Layer upon layer of biscuit-coloured dunes, hundreds of metres high, crowded the coastline like an anxious audience clamouring to view the ice-blue performance of the giant Atlantic rollers. The breakers roared their approval unceasingly.

We drove down to the beach opposite Mercury Island, so called because it quivers at the assault of storm waves. A quick hike over a rocky ridge brought us to a bay with the wreck of a large freighter high up the beach, seemingly crewed by hundreds of Cape fur seals.

Further north the dunes were littered with what seemed at first glance to be black confetti. The pieces grew bigger as we travelled up the coast until we came to a fantastically twisted metal sculpture strung about with huge, half-buried anchor chains.

The bits were the remains of *United Trader*, a 6,000-ton freighter which, in 1974, fetched up on the rocks carrying 700 tons of explosives. It had come from Sweden and was bound for Bombay, but the South African Defence Force had their suspicions about a more disturbing destination and blew it up. According to Volker, a helicopter more than a kilometre away was nearly downed by the blast.

Back at Saddle Hill North that evening I announced I was going for a walk up the high dunes.

'Don't shit up there,' warned Volker.

'Why not?' I asked, startled at this frank invasion of my privacy.

'Brown hyenas,' he grinned. 'They're okay generally, but if you shit they'll go for you. The top dog must have the strongest smell. They'll take it as a challenge.'

I set off, scanning for the tracks of turd-sensitive hyenas, but found only those of gemsbok and jackals. In the fading light the wind rills made beautiful regular patterns in the fine sand. I stopped beside a curious break in the symmetry – a set of curved grooves forming a rose shape. Dropping to my knees next to it, I snapped the pattern from close up with a 20 mm lens, then whistled my way back to base for supper. Several weeks later – browsing through a book on the Namib – I discovered what I'd photographed: the hiding place of a side-winding desert adder.

Next morning I strolled down to the beach. It glittered in the slanting rays and I scooped up a handful of pebbles. They were a paintbox of colours and some were as clear as glass. Maybe they were diamonds. My hand hovered at the mouth of my pocket for a moment, then I threw them back. They looked better on that wild, forgotten shore. After breakfast we packed up and backtracked over the massed sand mountains towards the dune streets. Volker was right: our earlier tracks had been obliterated.

On a particularly mean dune we looked back to see Gavin's Pajero, the next vehicle in the line, hopelessly stuck just below the crest, sand spiralling out from all four mudguards.

'Perone, Perone,' the radio crackled. 'You seem to be up to your chassis in trouble. Could you confirm? Over.'

'You can see that, dammit Chris,' came the answer, polite conventions abandoned.

'Now, now, Perone,' came the reply. 'Don't be rude. We'll just sit here at the back of the queue and watch you dig out. Okay? Over.'

Volker had other plans, however. He gunned down the dune, hitched a tow to the Pajero and hauled it backwards out of the soft sand. Gavin zigzagged up a safer route and ended on top of the dune in time to see Chris up to his chassis in the same sand trap.

'Carolin. Carolin. Do you read me? Over.' The voice was silky smooth.

'Yeah, okay Gavin, give it to me,' came the disgruntled reply.

'I suggest you do a bit of digging yourself. Then eat a good few mouthfuls of humble pie with as much sand in it as possible. Just an idea. Use it. Don't use it. Over.'

Back in the dune streets a stiff breeze was blowing – as it does on around two hundred days a year in these parts – moving grains and blurring the dune edges. In the mono-coloured dunes the only way to see walls and drop-offs is by detecting their sharp sand edges and these had merged into the ever-moving surface. It was like driving over a brown, undulating blanket thrown over countless hidden dangers.

'This can cause motion sickness,' said Volker, hurtling over the blur seemingly unconcerned.

'How the heck do you drive?' I wanted to know.

'Oh, you get used to it. When you see the horizon rising you know there's a dune coming, when the horizon drops you know you're cresting one. Occasionally the wind will gouge sand walls. If you hit one of those you're in big trouble.'

I didn't find the last bit of information consoling.

What we finally did hit was not a wall but a tar road – and a collective downer. It was a mighty long way back to Springbok and the comfort of a fur-lined bed. After the Namib's wild dunes, it looked set to be dead boring. The road flattened out and all that could be heard was the thrum of recently re-inflated tyres. Some guys cracked a beer, others contracted their spines into the comfortable seats and introspected. After about an hour the radio crackled:

'Carolin, Carolin, Carolin. Do you read me? Over.'

'...'

'Carolin, Carolin ...'

'Hey Gavin,' came a whispered reply. 'He's asleep.'

'Thank #@$% for that. Over and out.'

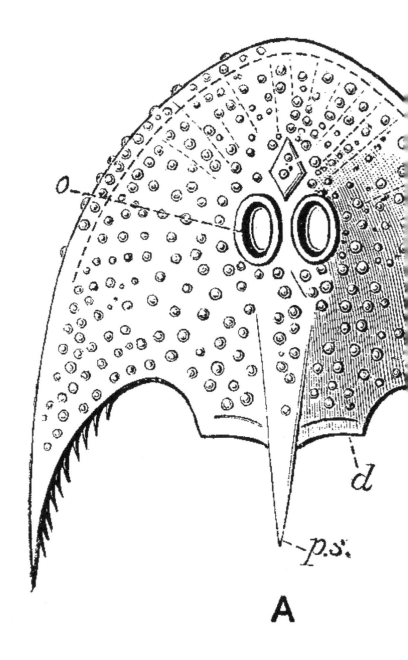

flying things

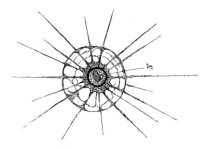

Superpilots

We were low on ducks. There were plenty of sacred ibises gliding in like great flying fish-hooks, and egrets in their thousands. But no ducks.

I was up to my underpants in rotten-smelling ooze, birdsquit and slime-covered reeds, staggering after Dalton Gibbs, who had a halo of rings round his blond hair and a thick wetsuit which made his progress a lot easier.

It's not that ducks were important, mind you – I wasn't even sure they migrated – but that's what I'd been led to expect we'd find in the Rondevlei reed beds.

'What do you use, mist nets or traps?' I'd enquired earlier as one of the team yanked the outboard's starter cord. 'No, nests,' Dalton had informed me. 'We're going hunting for chicks.'

The vlei's water comes by way of some down-at-the-heel Cape Town suburbs and probably a few squatter camps.

Puttering across the dull green water, Dalton had pondered the possibility of one of the three hippos which live in the vlei appearing suddenly through the reeds. Almost as an afterthought he mentioned the leeches.

Every time I looked up, a bird seemed to be squitting in my direction, and from time to time one of the chicks which could not escape our clutches vomited its last meal into the water. There are times when I wonder about my job, and this was one of them. If this is what bird ringers do, they are heroes.

Les Underhill, director of the University of Cape Town's Avian Demography Unit, had asked me along on the bird-ringing day. It involved a boat, a pack of eager ringers, a box full of various-sized rings and pliers to clamp the aluminium ferrules onto captured legs. Hence Dalton's halo – when you're in the goo it's the easiest place to keep the rings.

I asked the obvious question about ring size, with a vision of chicks' dragging leg-iron-sized rings or adult birds with shrivelled, blood-starved claws. After a certain age, Les explained, chicks legs don't grow any larger – in fact they shrink a bit. So it's okay to ring chicks. But first you had to get to them – through a hippo-patrolled, leech-infested swamp.

The object of all this, ultimately, was to help answer a few questions about these latter-day, long-distance, avian dinosaurs. Such as, what gets birds flapping half-way across the world? Where do they go? How do they navigate? How do they get their timing right? How do they optimise speed and energy during their awesome trans-planetary travels?

Avian migration is one of the great wonders of the biological world and our thrashing around in the reeds was, therefore, a noble pursuit, I kept telling myself, trying to ignore a strong suspicion that my legs were covered in leeches.

In relative terms, bird migration is a bit like taking a jumbo jet to the moon each year to pick up some in-flight meals and lay a few eggs. If you think I'm kidding, consider this: a turnstone – about as long as the spread of your hand – was recently captured four days after it was ringed 3,650 kilometres away. Another one, recovered fourteen days after being ringed, had flown some 7,000 kilometres.

But they're mere rock-hoppers compared to the Arctic tern, which can live for up to 30 years and is by far the world's long-distance migration champion. Beak to tailfeathers it's only about as long as a school ruler, but each year it makes a round trip between the polar regions of about 29,000 kilometres. Because of this it spends more of its life in daylight than any other species. It can even change its wing shape to achieve optimum performance in different winds and at various speeds. With those distances it probably needs to have a few tricks up its pinion feathers.

There are also spectacular stories about birds at high altitudes. A vulture was reported to have collided with a jet 11,300 metres above the Tanzanian plains, and hawks regularly pass over Panama on their way to South America at 6,000 metres – obviously, that way they avoid shipping congestion.

Even tiny songbirds migrate at around 600 metres, twittering to each other to ensure they all stay together. And they fly in great numbers. Researcher Ken Able, using radar over Louisiana in the United States, recorded traffic rates of songbirds at up to 200,000 birds an hour over a two-kilometre front.

Africa, it seems, is an avian favourite, and it's estimated that each year around five billion migrants cross the Mediterranean, though it's a pity that around one in five – about a billion birds – are killed annually by hunters on both sides of the Med. Why this feather-borne blizzard isn't common knowledge is because most birds migrate at night: at certain times of year there are millions of birds flying quietly through the black skies over our heads. They do this because of the cool, smooth, stable air that characterises the night-time boundary layer.

In the early 1940s, when Second World War technicians scanned the night skies with their recently invented radar systems, they saw mysterious moving dots and clouds which they named angels. Even today, many radar technicians simply describe these signals as 'anomalous propagation', unaware that they are countless flocks of birds.

Avian researchers, on the lookout for these 'angels' just after dark, say flocks tend to take off simultaneously, appearing as explosions of light on a radar screen.

There are certain givens that make such migrations possible: advantageous wing shapes, styles of flight – powered, bounding, undulating or gliding – and a range of sensory and navigation tools which would make a jet pilot weep. There are also instinctive cues which fire at appropriate times, plus an ability to map and take compass bearings.

But there is a growing consensus among ornithologists that prodigious feats of trans-planetary migration can't be passed off as mere instinct. Birds make decisions. It's a small-sounding conclusion with huge implications.

A pair of Cape robins in my garden have been flying their wings off raising a diederik cuckoo, which got its first growing spurt by eating the eggs of its siblings-to-be. The devoted pair must think they've spawned a terrible delinquent, who demands copious amounts of food and simply refuses to pattern on their robinness. But the affair raises some interesting questions about intelligence.

In mass migrations, first-time migrants will fly with more experienced birds which will guide them to their goal. But my clamorous cuckoo is different. At a certain moment it will take off alone and head for the forests of Central Africa on the first leg of its intra-Africa migration. It won't have been told how to do that by its foster parents, it won't have a crowd to show the way and will probably fly unerringly to its as-yet-unseen forest.

We don't really know how it does this, but ornithologists have drawn up a fascinating list of possibilities: the moon, the sun, the plane of polarised light from the setting sun, stars, wind, topography, olfactory cues ... Experiments in Australia suggest that the avian eye may even 'see' magnetic fields.

All those are mere tools: a bird also has to decide which cue offers the best shot in the prevailing conditions. On cloudy nights, for example, the stars may not be visible, or a magnetic storm may make magnetic cues unreliable. So the bird has to switch, and a decision like that doesn't rely on instinct or natural selection. It takes intelligence based on learning.

There are other examples of this intelligence. Weeks before the moment of migration, birds will become gluttons, building up fat – in some cases doubling their body weight – in preparation for the long flight. If they start this gorging too late they won't have enough energy to complete the flight, if they start too early they'll pop. Somehow they know, like we know when to start packing.

The clincher, however, is that birds have a phenomenal understanding of winds. If the wind direction is wrong or the blow too strong, they will wait. If the wind alters its favourable characteristics while they're in flight, they will search for more optimal layers. Soaring birds know just when to catch a thermal and when to leave it. During migrations they thermal-hop for hundreds of kilometres, dropping to roost when darkness drains thermal strength. In crosswinds they calculate the precise angle between flight direction and migratory direction, flying partly sideways if necessary to stay on course. Smart-alec humans have only just learned some of these techniques. Birds have been doing it for, oh, about 20 million years.

Quite how these small feathered creatures with pea-sized brains do all this raises an even more awkward question about mind.

Since the teachings of Freud and Jung we have associated

mind with brain, to the point where the two seem indistinguishable. But avian intelligence exceeds cranial capacity to such a degree that it seems legitimate to speculate about exactly where a bird keeps its mind.

The only obvious answer – which risks me being strung up by a pack of baying psychologists – is outside its body. Its brain (all of our brains, for that matter) may simply be analogous to a computer keyboard – it sends the signals but the hard drive is elsewhere. And external minds could overlap. The idea could offer a whole new dimension to telepathy, not to mention an interesting edge to migration studies. Imagine, telepathic terns beaming in the migrants. Dictionaries would have to redefine the term 'bird-brained'.

To get back to the ducks. If you take a ball-park figure, there are probably fewer ducks of any kind in Africa than people. We're stealing their habitats and they, obligingly, are dying off. At Rondevlei, however, they're probably simply being out-quacked by sacred ibises.

The Avian Demography Unit tends to interact with ducks when they ring them at birth and when someone sends a letter from some far-flung place, often many years later, reporting their death. Not infrequently the letters say, in effect: 'Yum, that was a nice bird. Please send us another one.'

That sort of thing's fine as another bit of data in the search for global migration patterns. But I wonder what the ducks think about it?

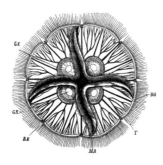

Our one big mistake

The discussion took place during a hike through the English countryside some time in the late eighteenth century.

Samuel Taylor Coleridge and William Wordsworth, both Romantic poets, were captivated by the far-flung voyages of captains William Bligh and James Cook. A mutinous crew aboard the *Bounty* had cast the former adrift, while Hawaiian islanders – whom he'd insulted – had killed the latter with a spear in the back.

The fate of these sea captains got Coleridge and Wordsworth thinking about irredeemable actions. They decided, while staring out over the beguiling hills of the Lake District, that at the core of all tragedy was the one mistake that placed a man beyond redemption.

Wordsworth went on to become poet laureate of Britain; Coleridge took far too much opium and died a wreck. They both wrote some beautiful poetry, but what concerns us here is Coleridge's *The Rime of the Ancient Mariner* – a poem brainstormed during the hike – and the notion of that 'one big mistake'. But more of this later.

Deep in the South Atlantic, aboard a vessel much more solid than the one in which Captain Cook had sailed these waters, I was watching a wandering albatross. It tracked us unerringly along invisible, wave-crafted updrafts, its great wings curved like some anti-gravity fighter from *Star Wars*. It never flapped once.

'That thing's incredible!' I exclaimed to birdman Warham Searle, who was sharing my perch at the stern of the RMS *St Helena* as it pitched through the rollers towards Tristan da Cunha.

'What's incredible is that it's still here,' he replied, gloomily. 'Seven years ago the waters round here were full of them. Each year I come here there are fewer and fewer.'

I stared at the huge, graceful bird silhouetted against the rising sun. Diving at incredible speed to mere centimetres above the wave troughs, it would then rocket up above deck height, curve into a tight, windward turn and hurtle down again, like a fighter plane dodging flak. Its ability to travel fast and accurately without seeming to move its body was almost supernatural.

'Where have they gone to?' I wanted to know.

'I think it's longline fishing,' said Warham. 'They're ripping the guts out of the oceans. Albatrosses and many other birds go for the baits and get hooked and drowned.'

I watched the beautiful bird for a while, then realised that Warham's comment demanded investigation.

Wandering albatrosses are huge, solitary, near-mythical birds which mate for life. With wingspans of nearly four metres they're the largest flying creatures on earth, and can spend many years without setting foot on land, seeming to revel in the fierce winds of the Roaring Forties. They are southern hemisphere, circumpolar birds, as are most of the other 20 albatross species. On their long, soaring wings they can travel at speeds of nearly 100 kilometres an hour and regularly undertake foraging trips of around 10,000 kilometres. They have lifespans similar to humans.

Back at home, John Cooper of the Avian Demography Unit at the University of Cape Town confirmed that albatross numbers are dropping fast – 20 species are considered endangered. In their last three generations, wandering albatross populations have gone down by 20 per cent. 'There are several reasons for this decline,' he said, 'but longline fishing is top of the list.'

Longlining is a relatively new technique which boomed following the international ban on 'wall-of-death' drift nets. Lines of up to 130 kilometres with baited hooks every few metres are spooled or 'shot' overboard and then reeled in several hours later. Current targets include tuna, hake, swordfish and Patagonian toothfish.

Because the line is generally shot from deck height, it's suspended in the air for some metres, then floats on the surface briefly before, hopefully, sinking. In the southern oceans thousands of petrels and albatrosses will hang round a single boat. If they grab a bait they can get hooked, pulled under and drowned.

It seemed essential to ask the local fishermen a few questions. Basil Lucas, of the Tuna Longline Association, met me aboard his boat in Cape Town Harbour. Bearded, friendly, slightly rotund and fish-smart, he's the sort of man you'd like to have along on a fishing trip. A giant spool on his boat's well deck held 80 kilometres of transparent line and several boxes at the stern contained thousands of business-like hooks.

'We catch maybe one or two birds on a trip,' he insisted, 'but that's it. We drop lines at night, we use bird-scaring tori lines and we don't throw offal overboard when we're shooting line.'

These were all good fishing practices. The implication was: look elsewhere. So I footed it round the quays to Irvin & Johnson, South Africa's biggest fishing outfit.

Barrie Rose met me in his architect-designed offices and was happy to share information. He's a bird lover, he confessed, masquerading as I&J's fishing-quota director.

Most of Irvin & Johnson's catches are hake, with one boat shooting for Patagonian toothfish. These, he pointed out, were both bottom fish and the longline technique – using fast-sinking line – reduced bird bycatch to a handful.

The smaller local outfits, he thought, were snagging more birds than they disclosed. But the real problem was Oriental fishing fleets and IUUs (illegal, unregulated and unreported fishing vessels). The Orientals – Japan, Taiwan, Korea and China – paid little heed to bycatches, he insisted, and IUUs – flag-of-convenience pirates registered in nondescript countries – obeyed no rules.

So the problem, evidently, wasn't in I&J's fish basket either. Next stop: the watchdogs.

I phoned Andrew Penney of Pisces Research and Management Consultants, which places observers on boats, and he also thought the local industry had a low bird bycatch.

'Maybe it adds up in the long term,' he told me. 'But in terms of how the fishermen see it, they don't catch many birds.'

But, it transpired, the local lads weren't all as squeaky clean as they seemed. At the tail end of a rather worrying discussion about longlining, Barry Watkins of Marine and Coastal Management slipped me a research paper headed *Seabird Bycatch by Tuna Longline Fisheries off Southern Africa*, by Peter Ryan and David Keith of the Percy FitzPatrick Institute, and Marcel Kroese of Marine and Coastal Management.

'I guess it's okay to give it to you,' he said. 'But you're not going to like it. Actually, it's frightening ... '

It was. The report was on the estimated mortality of seabirds within South Africa's 200-kilometre 'exclusive economic zone' (EEZ). It was compiled from observers on 13 commercial fishing boats, two being Japanese and the rest South African.

Multiplying the average bycatch for each 1,000 hooks by the 11 million hooks known to be shot in South African waters each year, the researchers estimated the annual bycatch in local waters to be between 19,000 and 30,000 birds a year – 70 per cent of them albatrosses.

Even though lines on the boats observed were shot at night, and despite the obvious fact that skippers with observers on board would tend to be more cautious, these numbers are scandalous.

Beyond South Africa's territorial waters the situation is much worse. There are around 3.5 million fishing vessels operating globally. These catch about 94 million tons of fish a year of which an estimated 27 million tons is discarded as 'bycatch'.

Not all of these boats are longliners. Not all the bycatch is birds – sharks, dolphins and turtles are also being dragged out – but a large proportion of it is. Around 100 million longline hooks in pursuit of tuna hit the water of the southern oceans every year. In blunt terms, it's an ecological disaster.

Bird bycatch has to do with fishing economics – and piracy. For every ton of fish caught by a registered fishing vessel it has been estimated that an IUU pirate lands ten tons – and they don't give a damn what else they kill. The pirates fish in daylight when birds abound, they dump offal while shooting line, they use huge lights at night which attract both fish and seabirds, and they're mowing down southern-ocean pelagic birds like aquatic chainsaws.

BirdLife International has reported that between legal and pirate boats, the Patagonian toothfish industry in the southern oceans kills around 145,000 seabirds a year, most of them white-chin petrels. But the tuna boats probably kill far more birds.

There are an estimated 15,000 sooty albatross pairs on earth. Royal albatrosses, perhaps 13,000. Wandering albatross, a mere 8,500 pairs. At present bird-catch rates, the end might come shockingly soon.

The IUU pirates hail from all nations but register their vessels, predominantly, in Panama, Honduras, Belize and Malta – countries prepared to license unsafe rustbuckets and which place no restrictions on their activities. Port Louis in Mauritius has been described as the pirate capital of illegal longline fishing in the southern oceans. Isofish, an Australian research organisation, estimates that some 30,000 tons of illegally caught toothfish pass through the port each year. Rumours have been circulating that Maputo may be the next port of call if Mauritian authorities crack down. There are fears that the species may soon be fished to extinction.

But pirates are not the only problem. China – new in the longline business – refuses to obey internationally set quotas and its huge fleet prowls the oceans without restraint. Taiwanese, Korean and Indonesian fleets take the same approach, and the Japanese have an insatiable demand for sushi and sashimi – raw tuna caught by a vast industry with albatross blood on its hooks.

The seemingly unstoppable destruction of albatrosses and other creatures has begun to ring alarm bells in high places. In an unprecedented article in *The Field* magazine, Prince Charles has pleaded for no-fish areas in albatross feeding grounds.

'No one who has seen an albatross on the wing is ever likely to forget the experience,' he writes. 'Just about the only way to do so is from the deck of a ship in one of the southern oceans, and that is how I came to know and love these magnificent creatures, while serving in the Royal Navy.'

In a World Day of Peace message Pope John Paul II made a

plea which could have been tailor-made for the fishing industry. 'Faced with the widespread destruction of the environment,' the pontiff declared, 'people everywhere are coming to understand that we cannot continue to use the goods of the earth as we have in the past. We have an ecological crisis,' he told the crowds. 'We need to take action.'

Bartholomew I, spiritual leader of the world's 300 million Orthodox Christians, was recently even more direct: 'To commit a crime against the natural world', he declared, 'is a sin.'

Among the many creatures which face extinction, the wandering albatross is becoming a *cause célèbre*. It's far more than merely the largest flying bird on earth; it has become a symbol of the great blue emptiness beyond every horizon. A lord of open skies and boundless seas.

In a way which the demise of the dodo threw human stewardship of the earth's creatures into question, the flight of the last wandering albatross may be the final judgement on that matter. Could it be the one mistake?

There's an eerie prescience in Coleridge's stirring poem *The Rime of the Ancient Mariner*. The ancient sea-dog's ship is lost among the ice floes of the southern oceans when salvation comes:

At length did come an Albatross,
Through the fog it came;
As if it had been a Christian soul,
We hailed it in God's name.

It ate the food it ne'er had eat,
And round and round it flew.
The ice did split with thunder-fit;
The helmsman steered us through!

But the Ancient Mariner shoots the bird, placing himself

beyond the pale and invoking terrible physical and spiritual tragedy.

And I had done a hellish thing,
And it would work 'em woe:
For all averred, I had killed the bird
That made the breeze to blow.

Ah! well a-day! what evil looks
Had I from old and young!
Instead of the cross, the Albatross
About my neck was hung.

All this seemed irredeemably gloomy, so I called Marcel Kroese, one of the authors of the South African report, hoping he'd cast some light. It wasn't a good idea.

'These birds are slow breeders – one or two chicks a year,' he told me, 'and they range over thousands of kilometres for food. The only way to prevent their destruction is to stop all longline fishing. And that's not going to happen. So, realistically speaking, I guess albatrosses have had it ...

'Is there anything else I can help you with?' he asked in the long silence which followed.

'Redemption?' I asked weakly, but I don't think he got it.

4

Parliament's poached penguin eggs and the Great Guano War

In June 2000 a bulk ore carrier named MV *Treasure*, owned by Good Faith Shipping, is holed by who knows what and sinks between Dassen and Robben islands off Cape Town.

On board are 1,344 tonnes of bunker oil, 56 tonnes of marine diesel and 64 tonnes of lube oil. Before long, hundreds of tonnes are in the sea and the shoreline looks a bit like the inside of a coal scuttle.

A few days later the slick hits Robben Island, then Dassen. Forget the irony of good faith and treasure: pray for the penguins.

This is old news, with resonances going back even further: *Kapodistrias*, *Apollo Sea* – bulk carriers which sank and spewed their black guts into the sea. *Treasure*, though, beats them all. It creates one of the world's worst coastal bird disasters followed by the world's most amazing seabird rescue operation.

Thousands of officials, professionals, housewives and schoolchildren are joined by international film stars, boxers, models and oil-skinned workers, grabbing, shooing, boxing, scrubbing and feeding bemused penguins.

Some 19,000 un-oiled birds are captured and transported to the safety of Cape Recife near Port Elizabeth, another 19,000 oiled birds are cleaned, tagged and, when the oil slick is dispersed, released.

This is twenty-first-century conservation at its best. It gives me a warm glow to have been part of it – until I meet University of Cape Town avian researcher Phil Whittington. He tells me that, until 1968, the South African Parliament had a breakfast special on penguin eggs. He mentions a Great Guano War and comments that in the last hundred years the African penguin population has crashed by about 90 per cent.

It's all rather depressing and takes some of the lustre out of our rescue operation. But what catches my attention is the guano war. Did grown men really kill each other over bird shit?

Well, it seems they did. From information I dig out of some obscure books at the University of Cape Town and a handful of government reports, a strange tale begins to emerge.

It really starts with the Incas, who discover they can grow bumper crops using the smelly white stuff they scrape off nearby Peruvian islets. The tradition persists despite the decimation of Inca culture by the Spanish conquistadores. In 1835 some Peruvian guano is brought to Britain. It lands up in the hands of Alexander Humboldt – after whom the Humboldt Current will later be named.

He discovers it to be rich in nitrogen and phosphates and shows it to be an outstanding fertiliser for wheat and turnips. Soon Liverpool merchants are daring the long passage to Peru and returning with holds full of 'white gold'. Guano becomes the world's first commercial fertiliser.

The African connection has to do with a rakish New York captain named Benjamin Morrell, who has a penchant for wandering the oceans. In 1828 he sails up the coast from the Cape to Angola, hunting seals. He eventually writes a book with the windy title *A Narrative of Four Voyages to the South Sea, North and South Pacific Ocean, Ethiopic and Southern Atlantic Ocean, Indian and Antarctic Ocean from the Year 1822 to 1831*.

An American critic calls Morrell 'a great navigator, a successful sealer and merchant, a voluminous and entertaining writer and a romantic liar'. But in his book is a sentence which will sign the death warrant of millions of seabirds, including African penguins.

Commenting on a visit to lonely Ichaboe Island off the Namibian coast, he writes: 'The surface of this island is covered with birds' manure to a depth of 25 feet.'

Morrell is interested in seals, not guano, so he sails off. But his book falls into the hands of Liverpool businessman Andrew Livingstone, who charters three small sailing ships to hunt down Ichaboe. Two fail but the third, a brig named *Ann* under Captain Farr, hits pay dirt in March 1843.

The sea conditions are awful, Farr has no materials to construct a landing stage, each longboat of guano has to hammer through Ichaboe's heavy surf and a southerly gale eventually parts the ship's anchor chains. But he sails back to Britain with a goodly load of white gold and Livingstone makes a fortune.

The businessman has a problem, though: nobody owns the island and he has to keep its whereabouts secret. He pays the ship's crew to shut up and sends them away on other vessels. But a crafty steward has a piece of paper on which is written '26 South 14 East' and which he sells to the highest bidder. The secret is out.

Before you can yell 'hoist the mainsail' the next guano hunter is sailing southwards – then another, then another. The steward is obviously doing good, unprincipled business.

The first ship to arrive is the *Douglas* under Captain Wade. He leaves a record: 'On first landing in November 1843 on the island which enjoyed for a time so odorous a celebrity, the place was literally alive with one mass of penguins and gannet. They were so tame that they would not move without compulsion. Thousands of eggs of the penguin, collected by the sailors, formed a savoury addition to their usual rations of salt meat.'

Captain Wade's men hack away at the guano cliffs in blissful isolation, but not for long. Soon other ships begin to appear over the horizon.

Before long Ichaboe is littered with spars, booms, topmasts, hawsers and sundry junk forming loading devices and crude shelters. So are the surrounding islands.

Within a month 20 ships are at anchor. Captain Wade takes possession of Ichoboe 'in the name of Her Majesty, Queen Victoria'. By early 1844 the bobbing fleet of guano hunters has swelled to around 100. By this time hundreds of men are camping ashore under flapping canvas. Claims are staked. Tempers flare.

Then a black southeaster hammers the fleet, causing collisions and forcing vessels to run for open sea. When the gale dies down and the ships beat their way back to Ichaboe, it's under new management. An Irish deserter from the Royal Navy named Ryan has convinced the temporarily marooned guano diggers to elect him president and has declared Ichaboe a republic. He demands £45 for use of the landing stage from each ship.

All hell breaks loose, aided by some smuggled liquor. No master or mate is allowed on the island; any officer attempting to land is pelted with dead penguins and threatened at knife-point.

Then a well-provisioned intruder into 'Bedlam Britannia' arrives, the American schooner *Emmeline*. Its master negotiates a deal with Ryan – provisions for guano – and its crew begins digging. The British are hopping mad. War is declared against the new republic.

Up to 2,000 men fight with picks and spades. The dead are hastily buried in the guano, unearthed by remorseless diggers and buried again in someone else's claim.

Finally a frigate, *Thunderbolt*, is called up from Cape Town and Ryan realises his little war is over. The marines land without opposition.

By January 1845 there are 6,000 men hacking away on the tiny island and 450 ships at anchor. No more incongruous sight has ever been beheld along that hostile, desert coast. 'It was a spectacle for the eye and mind,' remembered one Cape Town merchant, 'which probably has never had a parallel in the history of commerce.'

By May 1845 it's all over. About 300,000 tonnes of guano have been shipped to Britain at around £7 a tonne and most of the penguin islands round the Southern African coast have been scraped clean. Without guano in which to bury them, where are the corpses? There is no record of them.

What penguins remain have no burrows and they nest on the bare rock at the mercy of the elements and egg-eating kelp gulls. The islands are once again silent but for the cries of birds and the crashing of waves.

The Cape Government, bless it, then decides the islands need protection and in 1885 creates the Division of Government Guano Islands. This regulates the mining of the non-existent guano and formalises the theft of penguin eggs.

Years later – in the 1940s – the writer Lawrence Green visits Ichaboe and finds only a tiny penguin colony there. He laments that 'the dwindling of the penguins hits me in the stomach, for I know no finer breakfast than a penguin egg boiled for 20 minutes and served with butter, pepper and salt.'

This is a taste, it seems, shared by more people than is good for the beleaguered penguins. The tale is told in yellowing government records: state inspectors, it seems, are strangely obsessive in documenting the massive exercise in officially sanctioned looting.

A hundred years earlier it had been oil. A report in 1790 states that 'the Government sends every year a detachment into the Isle of Roben (Robben Island) to shoot mors and manchots, which are called at the Cape, penguins, from which they extract great quantities of oil'.

Then comes guano, then eggs – and the authorities appear to have counted each egg. In the peak years at the turn of the last century the average annual penguin crop from Dassen Island alone exceeds 450,000 eggs. The total documented haul on the 24 islands where African penguins nested between 1900 and 1930 is a staggering 13 million eggs. All duly recorded.

Penguin egg collecting is halted in 1969, though the parliamentary kitchen carries on the traditional penguin breakfasts under special dispensation for some time after that. And, of course, poaching continues.

African penguins are the only members of the penguin family which breed in Africa, and they pre-date human occupation by about 60 million years, appearing after the massive extinction of marine reptiles at the end of the Cretaceous period. They also pre-date seals, whales and dolphins. In the politics of life, they're elder statesmen.

Their nearest relatives are not puffins, which they resemble, but petrels and frigate birds. Penguins, however, exchanged air flight for water flight. Their Latin name, *Spheniscus demersus*, means 'plunging wedge' and plunge they can, reaching 20 kilometres an hour and diving to 130 metres when necessary.

But their common name, penguin, is derived from a Portuguese word meaning 'fat', and the relationship between humans and fat, flightless birds is not a good one. Ask the dodo.

So here's a question I'm left with. Was the disaster caused by the sinking of the *Treasure* the latest event in a long history of disinterest in the plight of penguins, or did the magnificent public response mark a new environmental awareness which will characterise the twenty-first century?

The answer, either way, is of great importance to embattled African penguins.

Angels of the night

The problem undoubtedly began with angels. If God was in heaven and humans were on earth, some monkish biblical illustrator must have reasoned, then the only way to commute between the two must be to fly. And the only flight he could imagine was the type birds did. So angels were drawn with light-filled frocks and white avian wings. Some time later pink cherubs with deftly concealed privates appeared, doing service around lovers.

However, the trouble came with Lucifer, God's fallen angel. He had to have wings too, being an angel, but not the nice feathery kind. One can imagine the scribe – probably a sixth-century Benedictine monk – scratching his bald pate and conjuring up an appropriate form for Beelzebub: a black, silent, night-shrouded, leather-winged, goat-horned embodiment of evil. It was to be very bad news for bats.

I was reflecting on this in the forest's inky darkness because it seemed preferable to the thought of being snapped in half by an enraged hippo which might soon run into our mist net. There were two other nets out across nearby game paths on Lake St Lucia's Nibela Peninsula – and hippos are nocturnal grazers. Merlin Tuttle, the head of Bat Conservation International, had his finger on the hair-thin net to monitor bat contact, and hippos didn't seem to worry him.

'Just dive through a gap that a hippo can't fit through,' he advised. 'They're fat and you're not.'

A slight jiggle of the net signalled the capture of what turned out to be an extremely irritated Egyptian fruit bat. Merlin disentangled it deftly as I made my way round the net to have a look.

What happened as I peered at the creature in his hand can best be described as an instant collapse of stereotypes: the bat was absolutely beautiful. It looked rather like a tiny, flying Jack Russell terrier, but with delicate radar ears, a long brown snout, puppy nostrils and the most intelligent eyes imaginable.

Merlin lowered it into a soft cloth bag to photograph later and I picked my way to another net, trying to remember what a hippo on the trot sounded like.

Peter Taylor of the Durban Bat Interest Group had just netted another flying mammal, this time an Angolan free-tail bat, which was much smaller and had a wrinkled snout rather like a minute bulldog. It put up a creditable fight for such a tiny creature before also being popped into a bag.

Bats had intruded in my relatively bat-free perception of the world some months earlier on Grand Comore. Someone mentioned there were bats on the nearby island of Anjouan with two-metre wingspans. Two metres! That was outrageous: I had to see them.

'Could I get to Anjouan?' I asked Omar, my guide for the day. Well, yes and no. He was sure there was a slow and costly boat from Moroni offering an overnight trip. It depended on the tides.

But also no, because we could get shot. The Comores, a tiny Indian Ocean archipelago, was having one of its identity crises and its four parts – Grand Comore, Moheli, Mayotte and Anjouan – all seemed to have proclaimed independence from each other. There were three factions of men with guns roaming Anjouan, so looking for Livingstone bats in the jungle right then wouldn't be a good idea. But, Omar said, he could show me the next best thing: flying foxes.

A hair-raising trip round the foot of a brooding volcano brought us to a towering takamaka tree. Hawk-sized flying foxes wrapped in their leaf-like wings hung from its branches. Dangling debonairly, often by one foot, they'd occasionally unwrap like Quality Street chocolates, stretch their jagged wings, stick out their little black snouts and peep at the world. They were way too high for us to see properly, but their dog-like muzzles were intriguing. So when I was invited to go bat hunting with Merlin Tuttle, I jumped at the chance.

Merlin is based in Austin, Texas, and is a population biologist turned bat friend.

'Do you know that nearly a quarter of all mammal species can fly?' he asked moments after we'd met. 'There are nearly a thousand species of bat: it's the most prolific mammalian order. Think about that.'

I did. And if *you* ever thought bats were yucky, disease-bearing bloodsuckers hard-wired to tangle in your hair, Merlin's the man to dispel your cherished myths. It will take him a few minutes to have you doing penance for previous attitudes towards the order Chiroptera (that's bats).

In Chinese culture bats are symbols of good luck, while in some countries they're simply a good meal. An Australian cookbook recommends flying-fox stew in spite of the 'strong and unpleasant smell which departs with the removal of the skin and wings'. Cooked with onions and herbs and boiled for a couple of hours, 'you would hardly know the flesh from pork.'

But apart from obviously half-starved outback types, European peoples have, for thousands of years, associated bats with graveyards, witches, the underworld and the Devil himself. For the brewing of some venal evil, Shakespeare had his witches in *Macbeth* stir in 'eye of a newt and toe of frog, wool of bat and tongue of dog'.

The real wool of bat, however, is the guff which seems to have stuck to these little mammals over the years.

Bats aren't blind, they wouldn't be so stupid as to tangle in your hair, there are only three species of vampire bats and they're all in South America. Bats are about as evil as Labradors, they've never been known to attack humans, they're far less likely to carry rabies than your Pekinese, and biologists such as Merlin – who've been poking around in bat caves for decades – have never become ill from bat droppings.

What bats do is far more interesting. Take the baobab. At a certain time of year, around sunset, baobabs will curl the white petals of their flowers upwards. Before long, straw-coloured or epauletted fruit bats will flutter in and sup delicately from the underside of the petals – clasping, all the while, the pollen-coated reproductive organs which hang as a convenient perch.

After an evening of boozing it up on nectar, the bats will be satiated and the baobabs will be pollinated. Without these pollinators, the baobab would die out, triggering a chain of linked extinctions of many other plants and animals.

Indeed, it has been estimated that up to 90 per cent of Africa's tropical forest trees and many savanna trees are pollinated by fruit bats. These include the many fig species, wild plums, water berries, wild pears, Cape ash, bitter almonds, cotton trees and sausage trees. They also do service for peaches, bananas, avocados, plantain, mangoes, guavas, breadfruit and dates.

Really evil little things, bats!

They are also vital forest re-seeders. Take wild figs. A single bat can take in and pass out around 60,000 seeds a night. Over a period of several nights bats may process nearly a ton of fruit from a single fig tree. Because flying with a tum full of fig consumes energy, bats let go between trees and in clearings – the most efficient way imaginable to re-seed cut-back forest areas. Birds, on the other hand, generally sit first before they poop, so their planting is done in much-contested soil below existing trees.

If only one per cent of the seeds dispersed by an average-sized tropical bat roost grew, it would mean something like 100,000 new trees a year. Anyone interested in rain-forest preservation should be praying to the god of bats to keep them safe, warm and well fed.

Not that it would interest them, but these busy little forest gardeners are causing a bit of a scientific storm among more recent mammals. Fruit bats have been around for about 50 million years (humans clock in at maybe three) and have been classed as megabats. Their insect-eating look-alikes are microbats. For scientific reasons rather too complex to go into, some hypothesise that megabats evolved from primates, while microbats evolved from a shrew-like, tree-living creature. They're about as related to each other as a tiger to a sea otter.

When bat-mum Kate Richardson hauled the fruit bat out of its bag it looked more like a lemur than a bat. She held it gently to prevent it struggling too much and, with a syringe, placed a droplet of apricot baby food on its tiny snout. It slurped it up with a long red tongue and smacked its lips.

After establishing that we were friendly and the apricot mush was in good supply, it settled back in her hand like a puppy on a pillow. It clearly knew how to train humans. I could swear I heard a little sigh of contentment. The Angolan free-tailed bat in the other bag was rather harder to please, and frowned at Kate comically until she produced a mealworm.

The family name for free-tailed bats is Molossidae, which comes from the Greek word *molossus*, a type of pug-nosed dog used by Greek shepherds in ancient times. Their strange little faces and large ears are part of the remarkable echolocation equipment which microbats use to detect their prey.

The battle between insects and bats has developed some of the world's finest bio-weaponry. Millions of years ago, before the appearance of bats, the night was safe for insects, many of which developed a nocturnal lifestyle. By using high-pitched clicks and buzzes, microbats developed a system of radar which, today, is millions of times more efficient than anything humans have yet produced.

Some insects were forced to move back to daylight activity, while others developed defence systems. One was the bat ear, found in certain moths, lacewings, praying mantis and perhaps some beetles. They can detect bat clicks and take avoiding action, either veering off, flying in wild loops or, if the bat has locked on to them with its 'feeding buzz', folding their wings and dropping to the ground.

Certain tiger moths go one better. At the final moment of the bat's attack, the moth blasts back streams of high-pitched clicks. It's rather like saying 'Boo!' to the bat, and very often confuses it enough to save the moth. Some bats, with stealth-bomber tactics, have retaliated by pitching their buzz at such a high frequency the moths can't hear it.

Fish-eating bats have tuned their echolocation to such a high degree that they could pinpoint a single human hair on the surface of a pond. They detect minnows swimming below the water surface, spearing them with a specially developed claw. Frog-eating bats know their favourite frogs by the songs they sing, and lock onto the sound unerringly. Juicy, fat croakers never stand a chance.

A good ear also helps with mothering. Bracken Cave in Texas houses between 20 and 40 million bats – the largest concentration of mammals in the world. Some 270 tons of bats roost in densities of about 5,000 a square metre. When the mothers go hunting, they leave their babies hooked to the roof amongst millions of peers, locating them later by their squeaks.

The real value of microbats, though, is the sheer volume of insects they eat: without them we'd simply be overrun. One colony of 20 million free-tailed bats can eat more than 100,000 kilograms of insects a night. That's the equivalent weight of around 20 elephants.

Many bats include mosquitoes in their diet: little brown bats, for example, can catch up to 1,200 mosquito-sized insects a night. This is something the citizens of the malaria-ridden town of Komatipoort evidently once realised. Earlier last century somebody erected two huge bat houses on railway property. One, measuring at least seven metres in length and three in height, must be the largest bat hotel in Africa.

This was a remarkable project, given that then, even more than now, bats were reviled, their caves dynamited and their roosts poisoned. Today nobody knows why or by whom the bat houses were erected. One still has a population of bats; the other is empty, its occupants having been recently poisoned.

The good news is that there are plans to restore the two 'batteries'. The really bad news is that bats are among the most vulnerable to extinction of any animal on earth. Most females produce only one baby each year. Others require up to five years to leave just two surviving offspring. In Europe, many bat populations are estimated to have declined by 90 per cent or more in the past 20 years and are now endangered.

On reflection, there may be a direct relationship with the increase in malaria and the decline of bat populations throughout Africa. More bats mean more forests, more fruit and fewer pests. A crash in bat populations could give us a foretaste of the hell so many people fear that bats represent.

After our captured bats had been fed and photographed, they looked rather contented. But they had things to do and places to go, so Kate carried them to the veranda. I shone a beam into the dark night and the bats lifted from her hands.

The light set their fur aglow and they seemed to dance into the sky. It was the nearest thing I'd ever seen to a flight of angels.

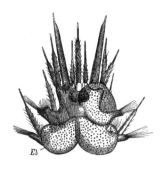

Balancing acts on the edge of extinction

'The mum's pulling out its feathers!' exclaimed Frederique de Ravel Koenig, hovering over an incredibly ugly, pink echo parakeet chick. She seemed close to tears. Soon there was a cluster of concerned humans arguing about a solution.

'Put it under a ring-necked parakeet,' suggested David Rodda from New Zealand.

'No. The chick will pattern on it and get the wrong messages.' This was Carl Jones. 'Put it back under the mum and watch. She must try to solve the problem first.

'I love radical approaches,' Carl grinned as we walked away. 'But when you've worked for years to get a single chick, a thing like feather plucking can be critical. There are so few of these birds.'

To be precise, there are an estimated 120 echo parakeets in the wild. Right now it's one of the rarest birds on earth.

Carl specialises in bringing back species from near-oblivion, which has proved to be good news for the Mauritian kestrel, the pink pigeon, the Rodrigues fruit bat, the Rodrigues warbler, a lizard, a fody and, of course, the echo parakeet.

His breeding centre at Rivière Noire in southern Mauritius, when I walked through the rather anonymous green gate, was as fecund as a honeymoon suite at one of the island's famous luxury hotels. A pair of tortoises were copulating, a rare Round Island lizard had just laid eggs, a fruit bat was nursing her batlet, a pair of kestrels were playing nooky in a tree, and parakeets were hard at it restoring their species. Only the pink pigeons looked sad: a bad case of in-breeding depression.

In *Song of the Dodo*, that extraordinary adventure in island biogeography, author David Quammen described Dr Carl Jones as 'a tall, sarcastic Welshman with a sheepdog haircut, a weakness for bad jokes and a manic devotion to native Mauritian wildlife, especially the birds'.

When I met him he'd lost the haircut, along (it seems) with the sarcasm and the bad jokes. But anyone outraged enough to spend 20 years of his life on a small, obscure island saving a single species must certainly be described as devoted. Indeed, he's something of a legend in his field, and though some orthodox scientists may be sceptical about his methods, his catalogue of successes is compelling evidence of a man with inordinate tenacity. And the list is growing:

The Mauritian kestrel: down to four individuals in 1974 and written off by the International Council for Bird Preservation as a lost cause, now up to 800 and its numbers are growing.

The pink pigeon: around ten birds in 1990 – a sneeze from extinction – but now at 430.

The echo parakeet: perhaps 12 birds in 1987 – only two or three females – now 120 in the wild.

The Rodrigues fody: from a few pairs in 1967 to around 1,300.

The Rodrigues warbler: from single figures to its present 150.

Low numbers lead to desperate situations and sometimes call for novel solutions. There's little room for failure. Take the matter of the hat. With the kestrel population dangerously small, Carl lavished his boundless energy and attention on every chick. In return the males – instead of falling for the females – patterned on him and fell in love with his hat.

They'd moon all over it. But it turned out they were doing more than mooning. So he let them, and artificially inseminated the females with the result. And it worked. Some 800 kestrels, if they knew how, should pay homage to Carl Jones's hat.

'Let's go up into the Black River Gorges Park,' he said suddenly. 'I'll show you some wild kestrels.'

Carl has far too much energy to sit still. He perches for a while, then paces, then watches a passing bird with a critical eye, talking all the while.

'Good stuff, all this,' he exclaimed as I got up to join him, throwing his arms planet-wide.

'It is, hey?' Since he wasn't talking about anything specific, I had to assume he meant everything in general.

He dug into his pocket and produced a limp yellow weaver. 'Trap 'em in my garden – non-endemics.' He held the dead bird at arm's length and a beautiful, speckled Mauritian kestrel swooped down from a tree and grabbed it.

'I used to be a falconer,' he commented by way of explanation as we set off for the gorges. 'You get to read birds.'

The park consists of 70 square kilometres of near-pristine woodland and forest. 'When the Bill to declare the park was before parliament, I brought the prime minister here to see the falcons,' he tells me with a wicked gleam. 'I got him to feed one by hand. In the parliamentary debate which followed, he informed the house you couldn't bring back falcons from extinction and give them nowhere to live. We got our park.'

A short way up a jeep track Carl whistled piercingly and yelled 'Come, come'. A falcon hurtled in at treetop level and braked spectacularly, dropping onto a branch above his head.

'That's a wild bird – and it comes when I call,' he enthused. 'Good stuff, all this, hey?' Another weaver appeared from his pocket, which the kestrel hawked in an elegant dive.

In different parts of the park, each time he whistled, a kestrel appeared. He knew each one, had worried over its parentage and its well-being: he's been coming into these valleys for years.

Carl also seems to have a bottomless pit of dead weavers in his pocket. 'They're in a plastic packet, but they didn't used to be,' he confessed. 'I had ten T-shirts and 20 pairs of identical khaki shorts. They washed well and that's all I wore. Then a girlfriend moved in and threw them all away. I said: "Hey, I've had some of those for ten years!" And she said: "Yes, that's the point … "'

The fact that Mauritius is the land of the dodo somehow makes Carl's work more poignant. Around five hundred years ago a Portuguese expedition under Alfonso de Albuquerque reached the island. There were no human inhabitants. For the next hundred years others of his countrymen dropped by but didn't stay. They never mentioned the dodo. Then, in 1598, a Dutch expedition arrived. They ate the native tortoises and they ate dodos. The latter, it seems, didn't taste so good, if one is to believe expedition commander Jacob van Neck:

'These birds want wings, in place of which are three or four blackish feathers. The tail consists of a few slender, curved feathers of a grey colour. We called them walckvögel (disgusting bird), for this reason: that the longer they were boiled, the tougher and more uneatable they became.'

Still they ate them. Dodos were slow, easy to catch and available (the name dodo seems to have come from a Dutch word meaning, roughly, fat-arse). When approached, they would hardly move out of the way. It was like picking apples.

The Dutch chewed their way through the island's tortoises, dodos, pigeons, turtledoves and grey parrots. Their appetite was ecologically unsupportable. A party marooned on Mauritius in 1667 provided the dodo's epitaph: 'When we held one by the leg he let out a cry, others came running forward to help the prisoner and were themselves caught.'

If another dodo was seen after that date, the fact was never recorded. Before long, 'dead as a dodo' entered the lexicon of cute expressions and the dodo became the legendary bird of extinction. Its disappearance was just a beginning of human-precipitated specicide.

Because islands are – by their nature – isolated, they give rise to extremely specialised creatures colonising small, island-specific habitats. But because islands are generally dots in the ocean, the number of species on them is relatively small. For this reason extinctions have been largely an island phenomenon.

According to ecologist Jared Diamond, since 1600 around 170 species and subspecies of bird have gone extinct, the most famous being the dodo. Of these, 155 species – more than 90 per cent – lived on islands. The Mascarene Islands (Réunion, Rodrigues and Mauritius) have lost 14 species.

Right now Carl's focusing on parrots and, as we wandered the Black River Gorges whistling up kestrels, I was treated to a Jonesian discourse.

'Parrots are avian primates. Think about it: they're arboreal, have sophisticated social behaviour, live in complex environments and spend a long time rearing their young. They're extraordinarily intelligent.

'We put them in a cage in the living room, give them no stimulation, and wonder why they display aberrant behaviour like pulling out their feathers. They're deprived, they don't develop emotionally. People have been stuffing them up for centuries. Cages are deprivation chambers – you've only got half a parrot in

Balancing acts on the edge of extinction

there. But – despite all that – they're still wonderful companions.

'We have to realise that birds live in a far more three-dimensional world than we do. They communicate by body language, by flight, even by where they perch. Ever thought why parrots can be taught to speak? It's because they're social and have a huge vocalisation range. They've developed a primitive form of language. They can learn, innovate.

'If we cage them, we have to also saturate them with stimulation. And they're highly social: they need other creatures around. We're only starting to understand what's going on there.'

Carl is pretty definite the answers about avian intelligence aren't going to come from conventional scientists. The narrowness of university training and the limitedness of the questions scientists are trained to ask clearly irritate him. 'Science is creative,' he grumbled, 'but we're taught to be so ... reductionist.'

He confesses a delight in taking young PhD students onto his programme and 'turning them into decent researchers after clearing out all the crap they've learned at university.

'In the field you have to understand the whole, functional animal. Understanding comes from empathy, not statistics and scientific method. You must start off broad, be intuitive, have gut feeling. You can bring in the scientists later. I'm a Lorenzian – you know, Konrad Lorenz? I can't stand back from what I study; I'm part of the equation. The world's such a wonderful place. It makes me emotional.'

We sit under a tree for a while, surveying the richly forested gorge rimmed with jagged volcanic teeth, and I ask Carl the question I've been dying to ask all day: 'If a species is so close to extinction, why bother to save it?'

'Okay,' he says after a few moments' reflection. 'I'll try for an answer. The most wonderful thing on this earth is its diversity. It'd be a hugely impoverished place with nothing but concrete and us. We'd be like birds in cages. People need their particular environments, but when they're exposed to the wonders of the wild they grow up with a richer spiritual sense, a different take on the world. Wild places help us develop depth in our lives.

'What we're doing here in Mauritius is learning to put back pieces of the jigsaw puzzle. We've got some islands – whole islands just out there, Round Island's one of them – which we're returning to pristine places. We can restore whole ecosystems if we try.

'We're also developing transferable techniques. I really believe it's not that hard to bring back species. There are very few variables: food, predators, nest sites, weather, habitat destruction. We can adjust most of these, intervene. There are critically endangered birds out there that could be saved for want of a few nest boxes.'

His best example is the Mauritian kestrel. Because breeding restarted from so few individuals, breeding pairs are now more closely related than siblings. Geneticists claimed in-breeding depression would prevent them from adapting to new environments. But they're flourishing. Now they're even breeding in towns.

'When the echo parakeets are safe, what then?' I wanted to know. 'New Zealand kias – and ravens,' Carl says, flipping a weaver carcass to a hawk still too shy to take it from his hand. 'The basics are in place here. I want to move on, set up a place in Wales and study these birds. Kias are the most intelligent of parrots, ravens are legendary. I want to give them massively stimulated environments, see what we get. I want to find out – and show the world – the extent of the avian mind.'

We make our way back to the Mauritian Wildlife Captive Breeding Centre. At the green gate, while I prepare to leave, Carl checks the trees for a while, hand in his pocket. Then he looks at me, grins and says just what I expected him to say: 'Good stuff, all this, hey?'

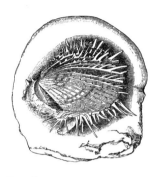

Sinking gannets and the mystery of a feather

Some 150 million years ago a single feather spiralled slowly to the depths of a Jurassic sea. There it was immediately covered by silt which would, eventually, become slate.

What is astounding is not merely that the feather was found, perfectly preserved, in the Solnhofen quarry in Bavaria, but that it was identical to those of modern birds. From the feather's construction it is clear: the creature from which it fell could fly. Was it a bird? Was it a dinosaur? Or was it both? Its discovery was to mark the beginning of a long and hot debate among palaeontologists.

The date of its discovery was 1860, a year after the publication of Darwin's *The Origin of Species*. One year later the quarry was to yield another treasure: a fossilised skeleton with imprints of feathers around it. Naturalist Thomas Huxley, Darwin's champion, seized upon the find as the perfect example of a transitional form between reptiles and birds. It was named *Archaeopteryx* and sold to the British Museum where it remained a puzzle and it was denounced as a fake. The press dubbed the fossil 'Piltdown chicken'.

There the matter rested until 1876, when the quarry yielded another *Archaeopteryx*. This one was perfectly preserved. The skull had teeth like those of a reptile. Its wings had three sharp, evidently movable, claws and it sported a long, saurian-like tail. All round the specimen was the clear imprint of feathers. They resembled those of modern birds; not just in number and arrangement, but also in their asymmetrical shape, with narrow outer vanes that could neatly cut the air. *Archaeopteryx* was no fraud, and four more of the creatures coaxed from the fine Solnhofen slate were to ensure its status.

But what was it? The feathers suggested bird, but the skeleton insisted dinosaur. Moreover, *Archaeopteryx* had no breastbone, which is the anchor point for the large pectoral muscles which power bird wings. So whatever *Archaeopteryx* was, it was not a great flyer: more likely a glider. But without doubt, it was some sort of fossil link between birds and dinosaurs.

Conventional wisdom hung there for nearly a century, with nothing turning up to fill the abyss between *Archaeopteryx* and later birds.

Then, in 1964, a Yale palaeontologist, John Ostrom, unearthed a three-metre creature he named *Deinonychus*, or 'terrible claw', related in form to the dinosaur group which included *Tyrannosaurus rex*. On careful investigation *Deinonychus* was nothing more than a scaled-up *Archaeopteryx*.

Thirty years later, in 1994, a Chinese farmer sold a fossil to a trumpet player, who gave it to the Chinese Academy of Sciences. It was named *Confuciusornis sanctus*, which means 'sacred Confucius bird'. A few years later, another bird-like fossil appeared from north-eastern China. This one looked like a small dinosaur, but along its back, neck and tail ran a fine ridge of lines which suggested feathers. It was named *Sinosauropteryx prima*, 'Chinese dragon feather'. It had lighter bones than *Archaeopteryx*, a shorter tail and undoubtedly better flying skills.

By then the connection was becoming clear. In the words of Ostrom: 'Dinosaurs did not become extinct. They live today as birds.'

From their reptilian beginnings, the ancestors of birds gradually developed traits that would aid flight skills: they gave up jaws with heavy teeth in favour of beaks, developed thin, hollow bones, perfected light, interlocking barbules on their feathers to scoop air, and souped up their metabolism to generate the energy to stay aloft. But what distinguishes them above all else is feathers.

These extraordinary protuberances – strong, light and beautiful – have raised much scientific debate. Did they develop for flight, or was flight the serendipitous by-product of a complex development in body-warming fur? The jury is still out on that. Complex they most certainly are. For their size and weight, feathers are immensely strong. They provide insulation, camouflage, protection, water repellence and display.

They also come in all shapes and sizes. There are body-covering contour feathers, flight feathers, pressure-sensitive filoplumes which sense the location of other feathers so they can be adjusted properly, fluffy down feathers for warmth, bristles which act rather like whiskers or eyelashes, special ear-covering feathers, and powder down feathers which shed a waxy powder of keratin to form a waterproof barrier for contour feathers.

A most extraordinary evolutionary invention, though, are barbules. Each flight feather has a shaft from which barbs extend. The barbs have their own barbules – smaller shafts – each of which has tiny hooks along its length. These tiny shafts interweave with barbules from the neighbouring barb and act like a zip, locking the barbs into place. When birds preen, part of what they're doing is re-zipping disengaged barbules – some birds even have a comb on their inner claws to do this. A well-zipped feather is warm, waterproof and beautiful.

While there are many birds which dazzle with their feathery iridescence, few birds match the sheer, air-cleaving line of a Cape gannet. With its long, pointed beak, yellow face and black wingtips, a flying gannet has the appearance of a winged arrow. Plunging into the sea from up to 30 metres, it conforms to that impression.

The greatest colonies of these daredevil divers are on islands along the Namibian coast, and it is on one of them, Ichaboe, that a strange feather malaise is taking place: gannets are sinking. The barbules on each affected feather fail to zip. The feathers become sticky and lose their insulation. The birds can't fly, and freeze in the icy Atlantic waters.

The problem was first noticed by Peter Bartlet, a researcher who has spent years on the Namibian islands, and came under the scientific scrutiny of zoologist Michelle du Toit.

Du Toit is a self-confessed wild island girl with a passion for lonely postings near roaring waves. She did an eight-month study of seal predation on Ichaboe, then returned for another six months to worry about gannets.

'When I arrived on the island Pete was the only person there,' she remembered. 'I doubled the human population. It's a pretty remote place. Our supplies came in only every six months, so if you ran out of something basic like toilet paper you were in big trouble.

'But for me the place was heaven. No time constraints, no people to hassle you, no phones, just you and thousands of birds. You learn penguinese and gannetese. You talk to penguins.

'Some days I didn't bother to wear clothes at all. Then I realised the birds watch everything you do. I had 100,000 pairs of eyes following me everywhere. You're never alone on Ichaboe.'

On her daily walkabouts, ringing chicks and counting nests, Du Toit came across increasing numbers of soaked gannets. It was possible that more gannets were dying from feather failure than from seal predation, starvation, diseases or any other causes.

Until pilchard stocks crashed in the 1970s, Ichaboe was one of the most important and densely packed coastal seabird breeding islands in the world and lies in the path of a powerful, nutrient-rich sea upwelling. It has one of only six Cape gannet breeding colonies in the world and possibly the largest. From 1956 to 1999, however, the population of gannets on Ichaboe plunged from 144,000 to 16,400. The area occupied by breeding birds in that time decreased by 90 per cent.

'Soaked gannets can be recognised by their inability to take off from the water,' Du Toit wrote in a report for the University of Cape Town's Avian Demography Unit. 'They are less buoyant and sit deeper in the water than healthy birds. If strong enough, they will swim to land, or even attempt to haul out on a boat.'

The plumage of stricken birds is wet and often with a yellow tinge. Feathers become sticky and smelly and they don't interlock, so the birds look scruffy. As far as Du Toit can establish, they all die eventually.

Why is this happening? Until further tests are done Du Toit is not sure. Fish oil dumped from hake trawlers is a possibility, but so much gunk is dumped into the sea it may prove to be a long, hard search for the culprits.

The cause, though, will probably turn out to be human. It takes our particular brand of intervention to interfere with a 150-million-year-old innovation which transformed a dinosaur into a bird.

swimming things

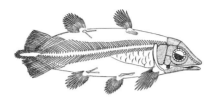

Everything but an eye-biter

'Their mouths are round like an O. They swim up to other fishes and suck their eyes out. Pop! Just like that.'

I winced. It sounded too horrible to contemplate. David Chitambo clearly relished the effect of his story, because his next one was about snout-sucking egg thieves.

We were sitting on the beach of the little fishing village at Cape Maclear in Malawi, watching the sun funnelling liquid gold into the afternoon sky above Mumbo Island. David is a fisherman and we were discussing cichlids, those colourful little fish for which Lake Malawi is famous.

There was reason to remember David's story the next day while snorkelling off Domwe Island. I was tail-up among a school of purplish cichlids which were nibbling algae off some rocks. A rather gaudy fish with a succulent pair of lips swam up to me and opened its mouth, displaying two rows of very businesslike teeth. It seemed to size up the single eye of my goggles, then flicked away into the grey.

Back home, fascinated by eye-biters, I began fossicking round in some scientific studies on cichlids. None mentioned the ocular specialists. So I called Tony Ribbink, the cichlid guru at the Institute for Aquatic Biodiversity in Grahamstown. His reply was brief: 'It's an error and a myth. No fish pluck eyes from living fish as a feeding strategy.'

By then, though, I was hooked on cichlids. Never mind the eye-biter story: the whole existence of this extraordinary family seemed to be a catalogue of errors, some awful, the rest a testimony to the wonder of serendipitous errors. I guess that statement requires an explanation.

Life has been around on earth for about three billion years, during which trillions of trillions of creatures have lived, some reproducing and dying every few days. That amounts to a big heap of trial and error.

We humans – present lords of the food chain – are the survivors of a long line of failures. We are apes, a group which almost went extinct 15 million years ago in competition with the better-designed monkeys. We are primates, a group of mammals which almost went extinct 45 million years ago in competition with the better-designed rodents. We are synapsid tetrapods, a group of reptiles which almost went extinct 200 million years ago in competition with the better-designed dinosaurs.

We are also descended from limbed fish, which almost vanished from earth 360 million years ago in competition with the better-designed ray-finned fish. We are chordates, a phylum that survived the Cambrian era 500 million years ago by the skin of its teeth in competition with the brilliantly successful anthropods. Our ecological success came against humbling odds.

It has made humans quite versatile. We have become what the geneticist Richard Dawkins terms survival machines, spectacularly successful creatures capable of existing in almost any ecological niche in the world. Right now there are some six billion humans, amounting to around 300 million tons of biomass. That's impressive.

The unravelling of the genome – the DNA chemical helix that makes all life possible – has shown us, surprisingly, that error is the engine of adaptability and perfection is the kiss of death.

If you think of the genetic coding as a book, it would have 23 chapters called chromosomes and more than a billion letters known as bases. That's as long as 800 Bibles. Genes are brilliant at replicating themselves over millions of years, but if they replicated perfectly we would still be squirrelling away at the manufacture of primeval slime.

Fortunately for evolution, letters sometimes go missing or the wrong letter is inserted. Paragraphs become duplicated. It's been calculated that humans accumulate about 100 mutations a generation. These are neither necessarily harmful nor beneficial, but without them we might still be fish or wombats. Or slime. Change equals adaptability.

You were wondering where the cichlids went? There's a connection. Lake Malawi is around two million years old and so, roughly, is its family Cichlidae. The African Great Lakes area was probably also the birthplace of *Homo sapiens*, which appeared about the time the Rift began widening its tear in the continent. So checking our progress against the little fish makes for an interesting comparison. Genetic trial and error buffed up our survival machines, but has left them largely unchanged since then. Cichlids took another road. They speciated. Breathtakingly.

From a single species of fish poured forth an evolutionary avalanche of adaptive species. Researchers guess that Lake Malawi contains around 1,000 species of cichlid, although almost every time they probe another species is found.

There are mud-biters that feed on detritus on the lake bottom; algae-scrapers that clean rocks; leaf-choppers; snail-crushers; snail-shellers; zooplankton-eaters; insect-eaters; prawn-eaters; fish-eaters; snout-engulfing pedophages which suck the eggs from the mouths of mouth-brooders. There are cleaners which remove parasites from other fish; stealth-trackers which hide on their sides under greenery, then pounce; egg-stealers; big-lipped beasties which suck food from rock crevices; and both left-handed and right-handed scale-biters which have their mouths on one side so they can better swipe the scales off living fish.

Some build sand castles under water to attract mates, others dig holes for the same purpose. The males of some mouth-brooders have egg-like spots on their anal fins. When a female with a mouthful of new-laid eggs makes a grab at the spots, the male squirts his sperm into her maw.

Some cichlids live among rocks, others above sand. Many live only at specific depths in the water column. These fish are so habitat specific that the entire population of a species may live in a single bay or on a particular side of a single island.

Most recognise each other by their patterns and colour, so snorkelling among cichlids can be dazzling. It's not surprising that Lake Malawi's cichlids are among the most popular occupants of fish tanks around the globe. Between them, the African Great Lakes contain more species of fish than any other body of freshwater on the planet. Around 99 per cent are endemic.

While our genes created a body which could carry us just about anywhere, cichlid genes – by trial and error – created a new species for every conceivable niche. They can speciate at an astonishing rate, with new species emerging within a human lifespan. By comparison, the coelacanth, that living fossil of a fish, has remained unchanged for 70 million years. Scientists are still puzzling about what causes the difference.

There is another error, however, that needs mentioning – not an error of gene replication, but of judgement. In May 1962, near Entebbe, in Uganda, a government official emptied a bucket of Nile perch into Lake Victoria. These fish breed fast and can grow to enormous size. It was thought that they would be a useful food source.

They were. At that time cichlids constituted about 80 per cent of the lake's biomass. Today, in many places, they come in at less than one per cent. Around 80 per cent of any netted catch is now perch, the other 20 per cent being tilapia and lake sardine. It is as if the food pyramid has been turned upside down.

In terms of biodiversity, the loss of the cichlid 'flock' to the hungry perch is staggering – it seems probable that about half of Lake Victoria's species have gone extinct or have population sizes so small that recovery seems highly unlikely. To survive, Nile perch are now eating each other, cannibalising their own young and doing very well on it.

Biologist Tony Ribbink has described the subsequent loss of more than 250 species of cichlid as 'the most dramatic example of human-induced vertebrate extinction in recorded history'.

The disappearance of big furry animals causes no end of public concern. Imagine the panic that would be caused by vast prides of lions in the Serengeti pursuing the last existing antelope. The drama going on below the waters of Lake Victoria is not much different. But beyond the circle of concerned ecologists the mass slaughter of cichlids by perch seems to be raising few eyebrows.

The effect on the area's ecology is already being felt. Swarms of lake flies – their larvae previously eaten by cichlids – have increased massively. Blooms of blue-green algae are spreading rapidly. Kingfishers are on the decline. Deforestation has escalated because the large perch cannot be sun dried and have to be smoked.

The trouble with extinction is that it is forever. We're losing species before we know they exist. Biologists have named around 1.5 million species of plant, animal, insect and micro-organisms, yet it's thought that the number of unknown species is anything between 3 million and 100 million.

A rough estimate puts the number of insects alive on the planet at about a million trillion. The number of bacteria in a single pinch of soil is around 10 billion. All that in a biosphere so thin that it can't be seen edgewise from an orbiting spacecraft.

Lake Malawi, like Lake Tanganyika, is still a treasure of biodiversity, a place of enormous ecological and evolutionary value. It has taken billions of microscopic errors and millions of years to turn the lake into a biological wonderland. It would take one human with a bucket – one massive error of judgement – to turn it into an environmental disaster.

Postscript: Cichlid speciation into every conceivable niche raises an interesting question about our particular niche. If Darwin and most modern biologists are correct, *Homo sapiens* is the product of chance mutations which were essential for our survival as a species in the niches in which we found ourselves.

All useless mutations were ruthlessly discarded: probably eaten by predators or bacteria. Humans, however, now live in nearly every available niche – in other words, across the whole planet.

That we developed a brain which could understand the nature and laws of the universe – and that it has seemingly retained that ability for thousands of years – means, in Darwinian terms, that it must be beneficial for our survival as a species.

The only reason can be that our future survival depends on comprehending the relationship between ourselves, our planet and the rest of the universe. And why? Ask any ecologist.

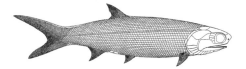

Lords of the sea

The ragged-tooth shark scanned Andy Cobb as he angled in towards it. But, as he matched his breathing to its gill movement and his fin rhythm to its tail strokes, it allowed him to swim alongside, accepting his movements as non-threatening. The two creatures from utterly different worlds swam side by side towards a pack of about 40 milling raggies, which parted gracefully to afford them free passage.

'You have to keep your cool with sharks,' Andy chuckled over coffee later. 'They're wonderful animals, totally adapted to their environment. They have so many finely tuned senses they can read you like a book.'

Andy's the grand old man of diving on the KwaZulu-Natal coast and his special favourite is Aliwal Shoal. When he first plunged among its population of migrating raggies, other divers declared him mad. Now the shoal has become a magnet for commercial dive operators. I'd asked him about shark sense and his answer was a sage one: 'Sharks have it and you need it.'

Durban had seemed a good place to start enquiring about sharks; not only because Andy runs his Eco Diving operation in the area, but also because it's the home of the Natal Sharks Board. The lurid display in the foyer of the board's offices in Amanzimtoti leaves you in no doubt about why the organisation was formed. It depicts a series of encounters between sharks and humans which – even as displays – are enough to scare you out of the sea forever.

'Black December', 1957, began when a 16-year-old amateur lifesaver was bitten by a shark at the South Coast resort of Karridine. Two days later a young girl was killed by a shark at Uvongo, followed by the death of a man at Margate.

December is holiday season in South Africa and widespread panic followed. On 30 December a girl was bitten in the surf off Margate and some ten days later a man was killed by a shark in Scottburgh. Natal's surf was cleared and people began leaving their hotels in droves. When more deaths from shark bites occurred the following Easter, holiday resorts along the coast emptied and many hotels went bankrupt.

In the anti-shark hysteria which followed, the tourism industry clamoured for some type of protection – the Navy even sent a minesweeper to depth-charge the Margate area (which killed eight sharks and attracted a whole lot more to devour their remains).

Barriers were constructed around swimming beaches but these were soon broken up by the surf. Nets were tried and, in 1964, the Natal Sharks Board was established to protect bathers by whatever methods possible. It is now one of the most prestigious centres of shark research in the world.

'Relatively speaking, people aren't shark food,' confirmed the board's executive research officer, Geremy Cliff, when I met him in his cool office packed with sharkabilia and weighty reports. 'But after Black December the shark became a monster in the public mind. When fishermen pulled one up and clubbed it to death, onlookers cheered.'

There seems to be a strange ambiguity at the board: people there study sharks, lecture about them, show wonderful films about them and regard them with the awe they deserve. But their job is to catch – and in most cases, kill – sharks.

'Gill nets are fishing devices and not barriers,' Cliff explained. 'They trap thousands of sharks and many other creatures as well – they're not selective. We net up to 14 species of shark – most not harmful to swimmers – as well as a bycatch of rays, dolphins and turtles. We now lift the nets during sardine runs because of this.

'Sharks are far more threatened by man than are threatening to him. But shark attacks are an emotive issue and they're costly to tourism. So, in the absence of any viable alternative to beach protection, the board has to kill sharks.'

Quite understandably, though, people have an ancient and legitimate fear about being eaten alive. Water is not our natural element and the thought of a silent, powerful killer hurtling up from below is enough to make even the stout-hearted nervous.

Of course sharks occasionally – very occasionally – do eat people. So do a number of other creatures if given the chance. Mostly sharks give humans a single bite and back off. But films like *Jaws* and *Blue Water, White Death* – together with lurid press reports – have led people to believe sharks will attack anything to appease their insatiable appetites. If this were true, not many swimmers would emerge unharmed: sharks possess such acute senses, have such wide distribution and occur in such large numbers that few people enter the sea without being detected. That's cold comfort for nervous swimmers.

In truth shark bites are incredibly rare. Fewer than 100 people a year worldwide are injured by sharks; less than 50 per cent of these contacts are fatal. By comparison, around 10,000 people were killed on South African roads last year; in the United States alone more than 300 people died of bee stings and nearly 200 from lightning strikes.

'Most "attacks" are either mistaken identity or provocation,' insisted marine biologist Dr Leonard Compagno when I tracked him down at the South African Museum in Cape Town. 'Sharks are inquisitive creatures but they don't have hands to test things with, so they mouth what interests them. The problem is we're so soft and thin-skinned.'

When I first phoned Compagno he wasn't too keen to see me. 'I don't like journalists,' he grumbled, 'but I'll give you half an hour.' As technical consultant to the film *Jaws*, he's become a bit wary about the way his research sometimes gets interpreted by sensation seekers. But he's a world expert on cartilaginous fishes and, among many other publications, has co-authored a guide to the sharks and rays of Southern Africa. He's fascinated by their behaviour.

'A shark may hit a surfboard and knock off the surfer but leave him alone. That's interesting. Or a diver will be digging abalone off a rock and not see a great white approaching. The shark may display and gape and get no response. So it will grab him by the leg and hold him for a bit. He'll flail and it'll hold a bit harder. But, generally, it won't complete its bite.

'Now if it hit a seal it would hit hard – the seal would look as if it had been through a slicer. It could do the same to the diver. But it will bite and not crunch. That's also interesting.

'Contacts like this with humans might be play, or purely accidental. But shark limits are different to our limits, so it necessarily gets logged as a shark attack.'

He's dead against coastal gill nets, which can kill more than 2,000 sharks a year. 'Nets are an unselective way of dealing with beach protection. It's rather like preventing a lion attack in a game park by shooting a few lions at random and then shooting a whole lot of other creatures as well – it's unscientific.'

Recent studies of sharks are raising questions about the popular notion of them as mindless killing machines fixated on humans as food. Both marine biologists and more experienced divers have become aware that sharks are complex, interesting and 'real' animals, not semi-mythical eating machines. They are, in fact, one of nature's most spectacular success stories.

Of the approximately 21,000 species of fish identified, only about 250 are sharks. Yet they are found in all the earth's oceans and range from tiny dogfish at around 15 centimetres to the whopping whale shark, at up to 18 metres the largest fish in the sea. They are also amazing swimmers, and have been known to migrate for thousands of kilometres.

Their form is so perfectly adapted to their environment that they have existed, virtually unchanged, since the Devonian geological period some 350 million years ago – the oldest-known vertebrates. With our five-million-year history we're relative newcomers to the planetary family.

Instead of a swim bladder, sharks have huge livers which provide buoyancy, allowing them to move from great depths to the surface with no ill effects. Their skin is a natural suit of armour covered with tooth-like scales called denticles. These are sharp enough to file off human skin on contact. During the recorded capture of a whale shark, a 12-gauge shotgun loaded with number two shot was fired into its back from a distance of 60 centimetres. The shot bounced off, leaving only a slight depression.

A shark's teeth are hard enough to cut metal – they're quite able to sever a chain or stainless-steel cable – and the bite of a two-metre dusky shark was measured at 60 kilograms across a single tooth or about three tons a square centimetre. As the gum tissue grows forward new teeth appear, allowing (in some cases) tooth replacement within a mere seven days. In their lifetime, sharks can produce up to 1,000 teeth.

Despite this impressive equipment, sharks are not greedy, though they can sometimes make gluttons of themselves in a feeding frenzy. Their large liver allows them to store nutrients for long-term use. Adults eat only about two or three times a week, males sometimes ceasing to feed during the mating season and gravid females (bearing young) perhaps not eating for their entire pregnancy. Sharks have been found to survive for up to 15 months without a morsel, though it's probably not a good idea to get in their range of vision after that.

But when sharks hunt they have no equals in the sea. Their battery of senses is so finely integrated that they are among the most successful predatory groups on earth. And they're fast: the pelagic mako is one of the speediest creatures in the sea, outswimming dolphins, sea lions and even swordfish.

Using their inner ears and two lateral lines of auditory sensors down their bodies, sharks can hear sounds hundreds of metres away. Their sense of smell is also acute, enabling them to detect around one part in a million. They've often been observed unerringly following odours for hundreds of metres through turbulent seas. Although their vision is not as sharp as ours, their eyes are able to detect objects and movement in extremely low light, which is why they are active night feeders.

As if these senses were not enough, sharks have electro-receptors which can detect weak electrical fields caused by muscular action. This allows them to sense movement and possibly to navigate along electrical gradients caused by the movement of sea water through the earth's magnetic field.

Studying shark behaviour in the Bahamas, researcher Erich Ritter strapped heart-rate meters to students and found that sharks tended to orient round those who had the fastest heart beats. Among other things, it proved the value of staying cool in their presence.

The most notorious of all sharks is undoubtedly the great white.

It's a huge creature with a blunt, conical snout and a large, triangular mouth full of saw-edged teeth. At full maturity it can reach more than seven metres. A great white was the star of *Jaws* and is the creature most people worry about when they contemplate the murky depths beneath their flailing legs. They've also been known to leap right into a boat in pursuit of a hooked fish. But, despite their fearsome reputation, great whites are not man-eaters.

'People are just not on the great white's menu,' emphasised Compagno. 'They are complex, inquisitive animals. If they bite people they tend to spit them out.'

Great whites are protected in South Africa and can be found round the coast from Angola to Mozambique. They're particularly abundant near seal islands in the False Bay area and off Gansbaai near Hermanus.

Cindy Thornhill of White Shark Ecoventures is one of seven operators based in Gansbaai and takes both thrill-seekers and serious ecotourists on day trips to Dyer Island. Her boat, *Lady T*, is a sturdy little nine-metre craft with a flimsy-looking shark-diving cage mounted on the transom. In pursuit of great whites we headed for Geyser Rock, then turned into the channel between the rock and Dyer Island, sometimes known as Shark Alley.

The sound and raw stench of around 20,000 Cape fur seals brawling, barking and defecating on Geyser Rock was overpowering; it hit me like a slap in the face. With so much food around the sharks regard it as a kind of fur-seal McDonald's. A chum of sardines and fish oil was trickled out to get their attention and a small dead shark attached to a line was thrown out as bait. Then we waited, the boat yawing in the swell with an awkward motion that soon had several people adding some chum of their own over the side.

When the white appeared its movement was casual – almost lazy. But its attention was not on our bait. The three-metre lead-grey creature grabbed a seal pup, then spat it out. It dived, arching its back as its elegant tail scythed the surface, then it mouthed the pup and spat again, like a dog with a tennis ball. Was the fare not up to standard? Was it satiated or was it just playing?

Finally it slid away, metres from our boat with that lazy motion which spoke of supreme power and confidence – a lord of the sea. The tatters of the uneaten seal bobbed grotesquely in the swell. After that the humpback whale breaching nearby seemed like a side-show.

Several hours later we were still waiting for a shark to take the bait, which would justify putting the cage in the water and kitting up for the Great Experience. But it was seal-pupping season and *Carcharodon carcharias* was out to lunch.

Further up the coast at Mossel Bay, *Getaway* magazine photojournalist Patrick Wagner had a wilder time cage diving with local operator Jimmy Eksteen.

'Suddenly the water seemed to explode around me and the white underparts and massive pectoral fins of a great white flashed into view. The shark was lodged against the steel bars of the cage and seemed to use the cage for support, wrapping its four-metre body round it and shaking the bait from side to side.

'The next minute will always be etched in my brain as one of the most awesome displays of power by a predator I have ever experienced. I was thrown around inside the cage like a rag doll, from side to side and lid to floor. All I could see was a sequence of white belly, green water, black-tipped pectoral fin and bashing dorsal fin. I was convinced the creature would tear the cage apart. Then it vanished as quickly as it appeared.'

There's been huge controversy about cage diving, with suggestions that chumming for and baiting sharks may get them to associate humans with food – with fatal consequences for swimmers, surfers, divers and spearfishermen. But, as Chris Fallows of African Shark Eco Charters pointed out when I put this to him, fishing boats have been gutting fish overboard for years with no seeming effect on shark-bite incidents. He's had lots of experience with sharks, with years of research and tagging behind him as well as ongoing work with trek-net fishermen to save netted sharks.

'It's also hard to imagine', he countered, 'how a dead shark dangled on a rope at Gansbaai could cause a great white to develop a taste for human flesh at some other part of the coast.'

A greater concern might be the flimsiness of the chicken-wire cage dangled before those steel-cutting jaws. But great whites seem to ignore human bait and do battle, instead, with more customary food hanging from the rope.

The people who seem to attract the most attention from sharks are not surf swimmers but scuba divers and spearfishermen, who can encounter the creatures nose to nose. It is they, most of all, who need to develop what Andy Cobb calls 'shark sense'. And this is where recent behavioural research is useful.

It's worth keeping in mind that sharks don't do unpredictable things, they don't attack without reason, humans are not their prey and they rarely make mistakes. That's why they've been around for millions of years. Despite their reputation, they are generally timid, though deep-sea species are more brazen than inshore types.

A feeding shark will first circle its prey warily, moving in only when apprehension is overcome by its desire to feed. Larger sharks tend to take fewer chances (which is probably why they're larger sharks) and it's the juveniles which need watching.

But most shark incidents with divers appear to be motivated by other than a desire to feed. Generally the issue is a territorial one, and the result is a slash wound similar to those found on sharks themselves, suggesting that such attacks are social 'warnings'.

Aggressive shark display is unmistakable, involving a laterally exaggerated swimming motion with the head and tail almost touching at times, back arching, head swinging, jaw gaping and a spiral gyration. These signals are common to most types of sharks, suggesting a form of ritualised behaviour with long-standing significance, and it is essential that divers recognise it. Such actions are both defensive posturing and a threat, and can be sparked off by harassment: diver pursuit, touching, cornering, unusual sounds, quick movements, swimming shoreward of them (especially among rocks) or the presence of a speared fish which the shark is denied.

Generally an attack occurs only when a human fails to indicate recognition of the threat ritual, and is usually lightning fast and very brief. Naturally, with increasing numbers of people using the sea for recreation (and spending longer times in or on the water), contacts with sharks will escalate.

Defence measures against attack have become increasingly sophisticated, including gas guns, explosive powerheads and electrical, chemical and acoustic deterrents. But nothing beats the ability to identify a shark, to know its behaviour and to learn what threatens it – basic shark sense.

However, shark nonsense – public phobia about sharks – coupled with both exploitation and callousness among commercial and sports fishermen, is causing the death of millions of sharks to every recorded shark bite. This is giving rise to extreme concern among people who understand their importance in the marine ecosystem.

Almost every day Japanese, Korean and Taiwanese trawlers, long-liners and factory ships are violating South African waters with impunity, decimating blue-fin tuna, removing thousands of tons of marlin and mutilating sharks for the shark-fin trade. Sharks are also falling prey to huge 'unintentional' bycatches in trawl nets.

'Nobody really knows what's going on out there,' complained

False Bay cage-diving operator Chris Fallows. 'But I see them violating our territorial zone all the time and nothing is done about it.'

Andy Cobb has noticed a recent dramatic drop in the sighting of raggies at Aliwal Shoal, Protea Banks and Southern Mozambique and he claims foreign fishing boats are disrupting their migration patterns. This has provoked him to call for an immediate halt to these practices and the protection, alongside the great white, of both whale and the ragged-tooth sharks.

'Sharks are far more valuable to us alive than dead,' he reflected, allowing his second cup of coffee to grow cold as his passion for the subject heated up. 'Quite apart from what they do for the ecosystem, sharks attract tourists. In the Bahamas it's been estimated that a live shark is worth US$20,000 and a dead shark $8. It doesn't take a rocket scientist to appreciate where their value lies.'

'Sharks are the ocean's top predators,' explained Leonard Compagno. 'Their decline would lead to overpopulation of other species – then disastrous die-offs. Sharks are being heavily exploited following the overfishing of other species – shark meat is mostly what you get in fish-and-chip shops in Britain these days.

'But shark populations are about to crash. Fish like tuna lay millions of eggs. If you put them under pressure they decline, then spring up when the pressure is released. Sharks breed very slowly, and harvesting them soon leads to overkill. It's a kind of meat-grinder effect – the populations can't come up again.

'Sharks are self-regulatory but the next layer of the marine population is not. The reduction of top predators is being referred to as an ecological time bomb with an ever-shortening fuse.

'The result could be what's known as a cascade extinction, where a whole ecosystem just collapses,' warned Compagno. 'If you bash a computer with a hammer, you shouldn't be surprised if it malfunctions. But that's what we're doing to the marine ecosystem. We need a whole new approach to the ocean – and sharks in particular.

'People enter their ocean world like a conquering army,' mused Compagno. 'We should do so with more humility and understanding.'

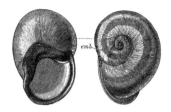

Snail hunting in the mother lake

'The life shape of a snail is an expanding circle turning through space and time. It's simple geometry, and it's so beautiful.'

I give Ellinor Michel a hard look. That's an impressive way to begin a discussion about gastropods.

Beyond her office, Lake Tanganyika glistens turquoise in the bright morning sun, a fuss of little fishing dories speckling its surface. Out on the veranda a few research students are clustered round two buckets, scrubbing snails with some defeated toothbrushes.

Ellinor detects my quizzical look and chuckles: 'When I was younger my heart wanted to be an artist but my head delighted in mathematics. So I ended up being an evolutionary biologist studying snails. It satisfied both tendencies.'

I'd met Ellinor by accident – she was looking for oxygen in

the same place I was enquiring after accommodation.

'Oxygen for what?' I'd asked.

'Snail diving,' she'd replied.

I thought she was having me on. But she wasn't.

Her snailing ground, Lake Tanganyika, is a huge, ancient and mysterious body of water formed when tectonic movements began tearing Africa apart along the Great Rift Valley (which is widening at about the speed your fingernails grow).

After Lake Baikal in Russia it's the second deepest (1,470 metres), second oldest (9–12 million years) and the second most diverse lake in the world. It holds nearly one litre in every five of the world's unfrozen fresh water, and is so deep that below 200 metres its waters have no oxygen and no animal or plant life.

It was once far deeper. Below its bed there are areas with up to six kilometres of sediment. A drilling project, soon to begin, hopes to find incredible layers of drowned life recorded in this mud – including *Homo sapiens*.

Tanganyika is thought to be the 'mother lake', providing the ancestors of fish, gastropods and other forms of life which colonised lakes Victoria and Malawi. At last count it was home to 250 endemic species of fish and around 60 endemic snails – a biologist's dream.

That explains Ellinor's interest in its crystal-clear waters. She reckons on another 30 snails soon to be described, and won't hazard a guess at the number of cichlid fish yet unknown to science.

Ellinor lectures at the University of Amsterdam, but for the past 14 years has been spending time at a remote little research station in Kigoma on Lake Tanganyika, where she teaches graduate students from Africa and America and carries out research on gastropods. She has a nose for big questions.

'Two generations ago my ancestors were missionaries,' she

explains. 'My father's a philosopher. So I grew up among people who asked questions which go to the root of existence: the meaning of life. We were always allowed to be intrigued by concepts.

'These days I ask the same questions as they did, I guess. I've just changed the methodology. And along the way I've made some pretty startling discoveries – the main one being that we're animals.'

Evolutionary biologists, I was to discover, tend to ask jaw-dropping questions. Among Ellinor's kind can be listed Charles Darwin, Alfred Russel Wallace, Jean-Baptiste de Lamarck, John Haldane, Julian Huxley and others: exciting people, when you can get your head around their conceptual timespans.

Gastropods, it seems, are an eminently respectable starting point for an evolutionary biologist. Ellinor picks up a tray of similar-looking snails lined up in orderly rows, largest to smallest. They seem much the same to me, but it turns out they're all different species.

'I'm not so interested in the creature inside,' she comments, prodding them thoughtfully. 'They're basically slugs. But their shells are a window on deep time. They were there at the edge – maybe 600 million years ago, at the beginning of hard bodies.

'And because they grow in a spiral, each shell is also a record of the creature's entire lifespan. Not many organisms do that: most get rid of their juvenile shape.'

Small aquatic things at the edge of time have had a large impact on evolutionary biology. Sharks' teeth come to mind, as do barnacles and a clam named *Trigonia*.

The teeth have to do with a man named Nicolaus Steno who, in 1677, became the Titular Bishop of Titiopolis (now part of Turkey). He was given some fossilised sharks' teeth from a quarry while he was dissecting a shark's head and found the teeth to be identical. This posed a problem: if God in his perfection had made the earth in seven days, why did He place sharks' teeth inside rocks? It also raised a more practical problem: how did solid bodies get inside other solid bodies?

Steno's genius was to recognise that all the troubling objects of geology were solids within solids, and an answer to how they got there might be a key to the unravelling of the earth's structure and history.

The sharks' teeth, he reasoned, were inside the rock because they had solidified before the rock enclosed them. Therefore the rocks must have been created after God's Creation; formed, rather, by deposits from rivers, lakes or oceans. And because fossils were often found on mountains, the processes that got them there spoke of great movements of the earth's crust and great gouts of time.

The Titular Bishop of Titiopolis is today, quite rightly, considered to be the father of modern geology.

Unlike the sharks' teeth, which advanced natural science, barnacles were so fascinating they retarded evolutionary thought for eight years. Charles Darwin, with all the material on hand to arrive at his epoch-changing theory of natural selection, was sidetracked by these potentially boring little rivets. In hindsight he said it was maybe the silliest thing he ever did.

This was a bit unfair to barnacles. Science writer David Quammen redeemed them in a charming essay, *Point of Attachment*, in which he explained Darwin's fascination.

A barnacle hatches from its egg as a tiny, six-legged beastie with one eye. After a few moults it transforms itself into an animal with three eyes, two shells, six pairs of legs and an inclination to give up the roving habits of its youth and settle down. It doesn't eat: its sole task is to find a suitable neighbourhood in which to attach itself, whereupon it transforms itself into a barnacle which never again moves.

Quammen speculated that what has become known as 'Darwin's delay' wasn't merely the fact that he was fascinated by barnacle reproduction (they have penises up to seven times their shell length), but that, being the reclusive homeboy he was, he identified with their tendency to hunker down and never move again.

However, prodded by an article sent to him by a young naturalist, Alfred Russel Wallace, which set out the principles of natural selection (and scared Darwin nearly out of his wits), the bearded Victorian got back on track and soon afterwards produced the path-breaking work *The Origin of Species*. Which brings me to another troubling little shell.

Trigonia is a distinctive clam which flourished with dinosaurs during the Mesozoic era a few hundred million years ago and then became extinct in the same mass die-off that wiped out the ruling reptiles. No *Trigonia* was ever found in the overlying Cenozoic strata and it therefore became a useful 'guide fossil'. When you found one you knew you were in Mesozoic rocks.

Then in 1802 a French naturalist, Pierre Peron, found a fresh shell washed up on an Australian beach. Some 25 years later two other naturalists named Quoy and Gaimard dredged up a living *Trigonia* in the Bass Strait between Australia and Tasmania.

The creature was duly killed in alcohol and sent to the Museum of Natural History in Paris. There it was first described by the institution's director, Jean-Baptiste de Lamarck, who used it to develop his pet evolutionary theory. Slight differences between the fossils and the living shell were his point of departure.

Lamarck – before Darwin – reasoned that living things descended from common ancestors and evolved (through challenges in their surroundings) into more and more complex creatures. The pressures to mutate, he said, came from the environment and there seemed to be an inbuilt tendency to ever greater perfection.

The giraffe had a long neck, he reasoned, because it had to reach for ever-higher leaves. In a few hundred million years the

Trigonia had managed only to develop a few extra ridges; obviously a slow learner.

Darwin – who ruminated over the *Trigonia* for 30 years – derived two separate theses about life on earth: he agreed that all organisms had probably descended from common ancestors, but insisted that chance variations and not environmental pressures drove the engine of natural selection. Many would fail, he said, some would make it and perfection had nothing to do with the process. *Trigonia* had evolved steadily; it's just that nobody had found the intervening stages.

Creationists such as James Parkinson had another take on the elusive clam. He delighted in what was known as the 'Cenozoic gap'. There had been a number of creations, he said, separated by sudden, worldwide paroxysms. God's creation was always perfect, needing no selection and no perfecting. It was just that there were a number of these creations.

So the little *Trigonia*, sunk in its gloomy bottle of alcohol, was taken to prove all things for all tendencies: that environment dictated evolution; that (by contrast) creature mutations were random and lived or died by survival of the fittest; and that God, in His infinite wisdom, created perfect creatures, eliminated them and created again. Because He could.

A heavy load for a lowly clam.

By the twenty-first century these versions of life's adventure were still bouncing around, though by then Darwin's argument was way ahead of the field.

The fact that at least 90 per cent of the creatures that ever existed have become extinct suggests that evolution is not the execution of some perfect, divine plan. There is, today, no doubt that evolutionary processes are at work. But some niggling questions remain.

Why, for instance, have some creatures – such as sharks and frogs – hardly changed at all in millions of years while gastropods and cichlids in the Great Lakes have speciated like starbursts? In Lake Tanganyika there are even left-handed and right-handed scale-biting fish (their mouths are twisted to one or other side and they eat scales from living fish). In almost every conceivable niche there is a creature specially adapted to fill it.

If survival of the fittest is the rule, why do less adapted species persist and seemingly thrive? If learning to fly provided birds with great new opportunities, why is the penguin very successfully turning itself back into a fish? If a cow developed such a complicated stomach for digesting grass, why is a horse, which eats the same stuff, so successful with a simple gut?

Is there really a gene which produces left-handed or right-handed scale-biters, or shifts the eyes of a flat fish round to the same side after birth?

Can we honestly expect chance and time to evoke the multiple modifications necessary to make the human eye? Even Darwin commented that 'the thought of it makes me shudder'.

Somewhere between the Creationists and the Neo-Darwinists maybe there is a 'third way', a Lamarckian tendency for forms to be what they are because of some built-in response almost amounting to – according to biologist Gordon Taylor – self-design. Or maybe some inherent blueprint?

Heretical? Well, so was the Titular Bishop of Titiopolis.

This brings me back to the lowly gastropod. In Lake Tanganyika, as elsewhere, they have developed thick shells because crabs delight in peeling and eating them. So, in a deadly 'arms race', crabs have developed larger, stronger pincers, which has prompted the development of more complex, knobbly snail shells which are even harder to crack. Well and good in evolutionary terms.

But does that explain the chance shape of a hundred different species? Why are there so many forms – seemingly far more than necessary?

In order to understand evolution, biologists need to figure out how particular body shape develops. There must be some 'laws of form' but, unhappily, they don't yet know what these are.

The little shell in Ellinor's hand, as she poses for a photograph after our discussion, is a beautiful thing – truly an expanding circle turning through space and time. The puzzle it poses, though, far exceeds its size.

'I keep asking myself: what are the motivating causes of diversity?' she comments as she drops the armoured slug back into a bucket of water. 'How much randomness does life need?'

A few hours later we're diving off a secluded beach south of Kigoma with handfuls of gastropods. Each one, duly numbered, we place back on a rock to continue with its life. The water is startlingly clear, and we can see the rehabilitated snails chewing away at the algae among thousands of their kind, quite unaware of their role in the meaning of life.

Ellinor, ever the teacher, spits out the snorkel and, with a twinkle in her eye, offers me an opinion entirely characteristic of an evolutionary biologist: 'We don't prove things in science. We only show them to be increasingly improbable.'

Then she dives back down into the mother lake with another handful of snails.

Neptune's little ponies

Sea horses flirt at dawn. Nobody knows why. Come early light the male will put up his head just so and his mate will swim up to him. Then they'll promenade and pirouette together, often holding tails and staring intently into each other's eyes.

After a few minutes of this coquettishness they'll separate and go about their daily business of hunting krill. As far as we know, they mate for life and do this little dance every day – except for the 20 million or so which get eaten as medicine each year.

I was watching a dancing pair in a corner tank of the Two Oceans Aquarium in Cape Town when my reverie was broken by the gasp of a young woman: 'They're real!'

'They're sea horses,' I said.

'Yes, I know that! But I thought they weren't, you know, actually real things – like unicorns … '

Well, she was from up-country. But I knew what she meant:

sea horses border on the surreal. I'd first paid attention to them in Port Nolloth where Marlene Gunter, who breeds them, put their oddity into perspective.

'The story goes that the Lord made the earth in six days, rested on the seventh and then, on the eighth day, put together a sea horse out of all the left-over parts.'

She had a point. They have a horse's head, a chameleon's independently mobile eyes and curly tail, the internal bone structure of a fish, the armour plating of a locust plus a kangaroo's brood pouch. Their elfin 'ears' are actually pectoral fins used for stabilisation and they suck prey into their long snout with an audible pop (they're voracious predators). Oh yes, and they can also change colour when mating or hiding out in the sea grass.

According to mythology, their job is to pull Neptune's chariot, though he must own some very large sea horses or be a very small king. They were clearly a puzzle to early taxonomists, who named their genus *Hippocampus*, derived from the Greek *hippos* meaning horse and *campus* meaning sea monster. They're related to equally odd pipefish, sea dragons and pipehorses.

Another quirk, in human terms, has to do with sex – but more about that later.

The semi-desert of South Africa's north-west coast was an unexpected place to find these strange little creatures. It was hot, dry and windy when I turned into Port Nolloth Sea Farms. One reason they're there is ever-caring Marlene, who calls them 'my babies' and recognises them individually. The other reason is the abundance of Atlantic krill, which they gulp down in relatively great quantities.

If their food isn't live they starve to death, which is why they are so difficult to keep in private fish tanks. Marlene's advice is: 'Don't even try it!'

She began with ten breeding pairs of the most endangered sea horse in the world – *Hippocampus capensis* from Knysna – which she obtained from marine biologist Jackie Lockyear at Rhodes University. She now has a whole room full of them.

'They do well on Atlantic krill,' Marlene chuckled as she scooped a fat fellow out of the tank for me to admire. 'My survival rate is around 90 per cent.'

Part of her success involves keeping an eagle eye on her little charges. 'Sometimes the males get too buoyant and float tummy up. I have to open their brood pouches and release the air. Or a bubble sticks to their tails and they flip upside down.'

Around sea horses, I discovered, talk soon turns to sex – in mammalian terms their style is extraordinary. The females, you see, have the equipment and the males get pregnant.

'They're monogamous,' said Jackie Lockyear when I called her to explain this oddity. 'They greet each other daily and engage in flirtatious dances. But courtship ballet preceding mating can last up to nine hours.

'At a certain point they'll drift up in a column of water, aligning themselves so the female can insert her ovipositor into the male's brood pouch and transfer her ripe eggs. Then the pouch is sealed and the eggs attach themselves to its walls where they're fertilised and oxygenated.'

Pregnant males, as you'd expect, can look pretty uncomfortable. I came across one at Two Oceans Aquarium hanging by its tail, head down, bloated and dull-eyed. He'll give birth – by pumping and jack-knifing – to around 200 tiny sea horses. As the last one pops out, the female will get the next mating dance going. As a result, the poor male will spend most of each summer pregnant. In some species up to 21 pregnancies a year have been recorded.

Despite this, more males, it seems, compete to get pregnant than females do to give away their eggs.

Sea horses have been around for some 40 million years, so I guess the males have learned to cope with the strains of being super-dads. Their real problems lie elsewhere and result from their being either cute or exotic, depending on your point of view.

'Cute' means people are enchanted by their appearance and buy them for home aquariums or expect to see them in public ones. Around a million are taken from the wild each year for this trade and, because they are difficult to keep, most probably die. A spin-off industry produces sea horses entombed in plastic and sold as key rings, paperweights and jewellery.

It's their exotic appeal, however, which is leading, inexorably, to their extinction. As early as 342 BC they were reported to have been used in Europe for medicinal purposes: Greek and, later, Roman writers claimed sea-horse ashes and oil of marjoram would cure baldness if rubbed into the head. If taken internally the concoction would, they said, cure leprosy, counter the poison of the sea-hare and neutralise the bite of a rabid dog.

In traditional Chinese medicine the little animal is credited with magical powers and has been used for thousands of years to treat anything from fatigue, asthma, cuts and grazes to incontinence, broken bones, heart disease, and delayed childbirth. In Taiwan they're taken as an aphrodisiac. You can buy sea horses dried or ground up in tablet form. Around one quarter of the world's population uses traditional Chinese medicine ...

Canadian researcher Amanda Vincent, working for Project Sea Horse, has estimated that about 45 tons of dried sea horses – that's around 20 million creatures – are consumed in the East each year. These are all plucked from their natural habitat by divers, resulting in a rapid decline in sea-horse populations worldwide – in some cases by up to 70 per cent. At least 39 nations are involved in the trade. In Hong Kong the demand is said to be 'limitless'. Almost all of the 32 species are now on the International Union for the Conservation of Nature's 'red list'.

Conservation, however, has its politics and its problems. Restocking depleted areas from tank-bred sea horses isn't good conservation – they could introduce diseases such as tuberculosis to already depleted populations. But breeding locally for the Asian market could simply increase demand and, along the way, introduce a taste for the Knysna variety. The answer, of course, is to preserve habitats.

In South Africa the threat to sea horses has more to do with habitat stress than to medicine. A single toxic spill into the Knysna Estuary could wipe out the entire population. All over the world, urban encroachment and pollution are damaging the seagrass beds, coral reefs, mangroves and estuarine environments which sea horses use as holdfasts and hunting grounds.

Thank heavens there are people who think small things like a single sea-horse population is a big issue. I have a sneaking suspicion that eco-warriors like Jackie – who spent five years of her life working on the Knysna sea horses – and Marlene are the planet's real heroes.

Someone once said that if you lift up the corner of anything you'll find it hitched to the rest of the universe. Conversely, if we lose a single species – say the Knysna sea horses – we might be doing greater damage to that universe than most of us realise. Nothing provable, mind you. Just an intuition.

If we don't get it right, sea horses may soon become merely mythological. After 40 million years of biological fine-tuning that would, to say the least, be a crying shame. In the meantime, quite oblivious of the fact that their future is in our hands, Neptune's graceful little ponies will continue to hold tails, gaze into each other's eyes and dance in the gentle light of dawn.

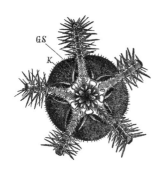

Monster hunting in the Okavango

Consider the louse. *Chonopeltis liversedgei* to be precise. It has eight legs, scimitar claws, two abdomens, an enlarged head with two huge suckers which look like eyes, and a set of bolt cutters that can do real damage. In appearance it's somewhere between a masked ballet dancer and a monster from the planet Zog.

But wait. It's also a romantic little creature which does a cute courtship dance before it makes love. And without it Okavango bottlenose fish would probably have gone off their rockers from itchy gills long ago.

This information probably isn't going to change your life. But, hey, it's good news for bottlenoses. There are only three lice on each fish: a female in each set of gills and a male who flutters concernedly between them. If another louse dares to shimmy up to the gills, it'll be attacked by the resident gill warriors and sent packing, sometimes with bits of its anatomy missing. Happy fish: no gills with crowds of free riders.

The louse's diet, though, sounds pretty yukky: gill mucus. They slurp it up with a Velcro-like tongue and snip it into edible portions with their on-board bolt cutters.

These somewhat bizarre facts come courtesy of a dedicated band of parasite hunters camped on the banks of the Okavango Panhandle up near the Namibian border in Botswana. I'm sitting in their caravan – no beds, just desks and equipment – squinting through a powerful microscope at Mr or Ms *Chonopeltis liversedgei* – it's hard to tell them apart.

My stool has just been vacated by Jo van As, an energetic professor and parasitologist from the University of the Free State. Together with his wife, Liesl, and colleague Linda Basson, they've set up camp in the Okavango – the largest inland delta in the world – with a group of research students who, at this moment, are up to their wrists in snail slime and fish guts. In parasitology, it seems, before the glory there's the rough stuff.

There are monsters out in the swamps you wouldn't imagine in your wildest dreams or nightmares. It all depends on your scale of magnification. I'm not talking crocodiles, of which there are an unnerving number.

As I squint through the eyepieces I'm treated to a little lecture. 'Throughout the ages parasites have had an appallingly bad press,' Jo tells me, prodding the louse with a small paintbrush to induce it to turn over. 'The reason is that humans are newcomers on this planet. We haven't had enough time to co-evolve with our parasites. They take us out, which is pretty traumatic.

'For this reason our customs and religions are littered with parasite references, remedies and warnings. The problem goes back even further: we probably became hairless apes in an attempt to deal with external parasites. Can you remember your mother's horror when you came home from school one day with head lice?'

I sure can . . .

Fish are the oldest animals with backbones, pre-dating humans by around 400 million years. They watched the dinosaurs come and watched them go. So they've had way more time than we have to get comfortable with their parasites.

Earlier that morning we'd been fishing, and the results of our efforts are cruising round in plastic baths: just a few of the 88 fish species found in the Okavango system. It doesn't seem fair to tell them about the scalpel-wielding students. Some are beautiful; the catfish, on the other hand, look rather like aquatic cockroaches with cavernous mouths.

I'm offered a bit of useless advice by a student: 'Did you know there's a tiny parasitic catfish in South America that's attracted by human urine? It swims up your penis or rectum and feeds on blood.'

No, I didn't. And that's the end of peeing near rivers for me.

'Have a look at this,' says Linda, who's been peering into another microscope. 'These are my speciality.'

I move across from demon dancer to space angel. In the squidge of water is a silvery flying saucer far less than a millimetre across which bears the weighty name of *Trichodina labyrinthae*. It's scooting along with surprising agility, using fine, transparent wings all around its circumference.

These wacky watercraft are single-celled creatures which attach themselves to fish fins, skin and gills by adhesive discs, spin slowly (who knows why?) and filter-feed on passing tasties but not on their hosts.

To say the least, these little guys pose a problem for comfortable evolutionary science. The 'higher' creatures are dignified with the name 'vertebrates', having evolved a spine providing strength and mobility. Dissolving away the fleshy parts (evidently Coke works best), Jo and Linda have found that *Trichodina* also have flexible spinal columns. The only difference is they're circular – and beautiful.

I won't begin to debate the evolutionary implications of a

single-celled creature with a spine . . .

Along the way to survival, Okavango's fish parasites have developed some strange coping techniques. Take the roundworm which lives in the gills of pike. In an act of supreme sacrifice the female, when her larvae are ready to be born, simply explodes, releasing her young into the water. There they are eaten by plankton which are, in turn, eaten by young pike. The larvae, unharmed by all this gobbling, migrate through the pike's bloodstream to the gills (how do they navigate?) and the process is set for another round.

The Free State University crew are modest enough about the fruits of their annual expeditions. But it turns out that between them they've described for science no less than 83 new species and three new genera: enough to turn their European and American counterparts swamp-green with envy.

But the bad press parasites get irks them: 'I need to make a case for equal rights for parasites,' Jo chuckles later as we sip some of his lethal brandy in the crickety Delta night.

'Think about it. There are more species alive today than at any other time in the history of this planet. Yet this diversity represents maybe one per cent of the species that have ever lived. It took 100 million years for life to escape from a single-celled organism. Now look. It's all over the place. New niches constantly have to be found in ever-changing ecosystems. So it's not surprising that niches and habitats are found inside other living organisms. Anyway, we parasitise cows and lettuces. Where do you draw the line?'

One of the researchers, Nico Smit, tells me he's found parasites of parasites of parasites of parasites. Four deep and he's still counting.

Creatures can live in extraordinary habitats. Bacteria, those masters of extremes, thrive in boiling water, caustic soda lakes, strong acid, highly salty water, under enormous pressures and in the middle of rocks. They can tolerate high radioactivity, live without oxygen or sunlight and, failing all else, scoff oil, plastic, metals and toxins.

Woolly bear caterpillars, which chill it out in the high Arctic regions, spend as many as ten months frozen solid at around 50°C. Salamanders can do much the same. Even human sperm is routinely frozen in liquid nitrogen at temperatures of less than 196°C.

It's probable that many creatures, such as the bacterium *Helicobacter pylori*, which causes human stomach ulcers, ended up inside other creatures to escape that well-known pollutant: oxygen.

Way back at the beginning of life there was little or no oxygen in the atmosphere. Then, around three billion years ago, single-celled organisms named cyanobacteria arrived on the scene. They used sunlight to convert water and carbon dioxide into carbohydrates. Oxygen was the useless by-product.

Huge amounts were pumped into the atmosphere. The stuff was toxic to most life forms (and, in quantity, still is). The die-off must have been terrible. Those that survived the pollution evolved strategies to protect themselves, and one was to dive inside the bodies of other organisms which were better at handling oxygen.

When we realise we're each host to maybe 30,000 parasites right now, it makes you wonder what right we have to consider ourselves an individual. Our eyelashes alone are host to more odd beasties than is comfortable to contemplate. Parasites? Well, if we want to point fingers, we're living on the detritus of one of the greatest acts of pollution the earth has ever known: oxygen.

For all his enthusiasm, Jo's worried about the Delta. 'When we started our study, no information on the parasites or pathogens of Okavango fishes existed. You can't manage an ecosystem without that knowledge.

'Fish parasites are like black box recorders in aircraft. If you know how to read them you can uncover information about the past and determine where the system is going. Parasites won't normally kill a host, but if we disturb the water quality it can favour parasites and cause mass fish die-offs.

'Farmers are spraying insecticides and allowing cattle to encroach into seasonal floodplains. There's increased fishing, and mines and agriculture are claiming an ever-increasing share of water. All this can have a disastrous impact on the Delta.'

One of Jo's biggest concerns, though, is carp. The subject brings out his gloomy side, a sort of resigned, no-hope view of humanity that the black Okavango night merely deepens.

Carp are considered perfect for fish farming and there's talk of starting projects along the Okavango. The problem is that Delta creatures don't obey the normal predation pyramid with the predators at the top and the rest down below. Out there almost everything's a predator. And mostly they hunt by sight.

Carp are mud feeders. If they get into the system, their snuffling will turn the clear waters brown. The predators will starve and the uniqueness of the Delta will die with them.

By the second day I've become decidedly parasite friendly. They're fine little beasties; they have a life to live. Give them a break. I've even come to terms with the bug-grooming going on in my eyelashes. And so what if my stomach is full of hungry strangers? That's about the time I find out about the tongue-replacing parasite.

It's an isopod and doesn't live in the Delta: tidal pools are its stalking ground. When a fish is young the isopod pops into its mouth, devours its tongue, turns round, hooks onto the stump with its legs and becomes the fish's tongue. Rather efficiently. Except that it steals passing scraps of food.

Though I give the isopod full marks for ingenuity, this bit of information whacks the edge of my regard for parasites. A tongue-replacer! It's too horrifying to contemplate.

Cruising the Delta channels before dawn to check fish traps, then watching the crew peering at the bizarre contents of fish innards afterwards, I'm reminded of the White Queen's claim, in *Through the Looking Glass*: 'She sometimes believed as many as six impossible things before breakfast.'

There's a good parasitologist's motto in there somewhere.

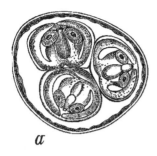

a

Wild orchestras of the night

With his hands cupped to his ears, mouth open and his thick, long ponytail, Marius Burger looked – just then – remarkably like a surprised bat-eared fox. On the nearby highway a truck sledge-hammered brute sound into the darkness, and a motorbike sliced the night like a white-hot machete. 'You hear that?' Burger demanded. 'Ping-ping-pingpingpingpingping – like a tiny marble dropped onto a hard surface? Dainty frog – *Cacosternum platys*. Over there. Hang on, I'll find it.'

He squelched through the soggy wetland, his long-suffering jeans clinging to his wet calves, and crouched – all bewhiskered attention – seeking the single note in the wild symphony of sound. Then he scooped up a tiny frog – no bigger than his fingernail – and sloshed back, grinning widely, to reveal the marble-dropper.

The scene was surreal. Our horizon was studded with rows of

lemon-coloured streetlights and city sound. To the west the dark silhouette of Devil's Peak stabbed through central Cape Town's eerie glow and, nearer at hand, marched the ghost-white barrier posts of the Kenilworth Racecourse. Gossamers of mist drifted over the wetland, sliding through our torch beams and hovering protectively over deep pools.

Around us, in the dark centre of the busy city, was a shrine to the Red Data Book of endangered species and a full orchestra of frog song – heard by few – whose contract with the planet could soon expire.

The frog we were hunting, though, was not the dainty *Cacos*. It was tiny – a micro frog – one of the smallest of its family in the world and critically endangered. Its Latin name, *Microbatrachella capensis*, is considerably longer than the frog itself. This speckle of biodiversity lives in the centre of the racecourse, with a few more around Betty's Bay and Cape Agulhas. Of course they have no idea they're that rare and, as I shivered in the dark, a declining number of males were bidding for a declining number of females in curious, scratchy tones.

Burger, I noticed, didn't seem to be cold. He's a long-time frogger – in fact he is National Co-ordinator of the Southern African Frog Atlas Project – so soggy nights are his natural habitat. Back in his Grahamstown days he used to be known as the Snake Man, and was blackballed from the local pet shop for feeding cuddly little hamsters to his two-metre python. These days, though, it's frogs that get him going. Frog atlassing is a business bordering on madness. The problem is that, unless you know how many of what frogs are where, you can't know whether they're doing well or declining. So these wacky herpetologists have divided South Africa into a grid with nearly 2,000 'cells' of 15 minutes latitude and 15 minutes longitude (about 27 kilometres by 23 kilometres) and hunt frogs in them with a GPS in one hand and a torch in the other. Of course, frogging is best done at night, when it's raining. Naturally. 'I start at around 6 pm and work the whole night,' Burger explained. 'Maybe I'll travel a thousand kilometres in three days covering a few cells. Sometimes it can take several hours to locate a single frog. It can mess with your social life a bit. People used to be really interested when I worked with snakes. Nowadays if I say, "Hello, I'm Marius, I like frogs," not many people can run with that line of conversation.'

He's had some tense moments poking round the backroads of the country. Farmers get jumpy, understandably, and arrive with bakkies and guns. Police have been called to arrest the 'stock-thief'. Being nocturnal and insatiably curious creatures by profession, froggers can get into some really odd scrapes. Frog hunting in the Central African Republic, Burger and some other biologists were cut off from their base by rebel forces. They had to backtrack into the jungle, travel down a river by dugout and jump the border into Cameroon, then talk their way onto a plane to Paris.

Then there's the Gabonese tart story. 'I was working in Libreville,' Burger explained as we headed for another racecourse pool. 'At some stage I picked up a fungal infection on my feet. I heard some frogs calling in a marsh in town. I could smell it wasn't a good marsh, but I was curious so I worked out a cunning plan. I left the hotel with my torch that night. But just outside a lady of the night said: "*Bonsoir, monsieur*. I am Nellie. You want some lurv tonight?" I tried to brush her off but she followed me. Then I thought: "Bugger, this, I'm going to carry on with my plan." I took out two condoms. She looked at me. I put one on each foot and took out a pair of latex gloves and also put them on my feet. Her eyes were like saucers. Then I said: "Bye, Nellie" and walked into the swamp. When I came out she was gone. But imagine the stories she told her friends about this crazy *mzungu* with condoms on his feet ...'

A frogger's most important piece of equipment, however, is not latex or a torch, but his ears. Male frogs are irrepressibly vocal wooers. They gwaak, blip, squeak, mew, squawk, bray, snore, quack, buzz, chirrup, and some – when threatened – scream horribly.

They were originally divided into species according to their appearance – generally from dead museum exhibits. But it's become increasingly clear that that isn't how frogs do it. A single species can come in many colours or patterns, but what's all-important is its call.

'A frog may go tik, tik, tik, gweek,' Burger explained. 'One part of the call may be aggression towards other males and another part a song to the female. Even in the gweek there are bits that she responds to and bits she ignores. She's virtually deaf to everything but the blip in the gweek. No blip means no breeding will be possible. And that really means a different species. So these days we are approaching frogs with a finer ear. By just looking at frogs as though they were a box of Smarties – and not listening to their conversations – we were denying them their wonderful complexity. For frogs it's more important what you say and how you say it than how you look. These are creatures of the night, remember. And the gift of the gab wins the woman.'

When frogs call they look as though they're blowing up a balloon. This is because they shout with their mouths closed, pumping air across their larynx between their lungs and a cavity that inflates under their lower jaw. Then they shove the air back into their lungs and do it all over again. This results in some very nifty sound effects and, during call sessions, an ungainly looking frog.

Male frogs get pretty excited about mating. Any frog that hops into sight when a male's calling into the night will be clasped, whether it's male or female. So frogs have even developed a 'let me go, you idiot' call that evokes instant release and, perhaps, slight froggy embarrassment.

Some frogs virtually glue themselves to the backs of females;

others will cling on for months to make sure they're 'The One'. When eggs appear, some species clasp each other in large mating orgy balls.

There are, at the last count, 110 species of frog in South Africa and each has a unique call. Burger records them, then runs the sounds through a computer that selects out individual frog calls, creating an unmistakable visual imprint.

'Basically, I'm barcoding frogs,' he chuckled. 'It's the best way to sort out species. There will be a good number of new species described before we've finished this.'

Although the atlassing project gives the impression there are more frogs around than we imagined – simply because more are being counted – frog numbers are declining.

At a World Conservation Union conference about 15 years ago herpetologists compared notes and realised that mass mortalities were occurring all over the world. In pristine rainforests you'd see dead frogs lying all over the place. Nobody had seen such a die-off before.

A task force was set up to investigate and it has come up with some tentative answers. These include the usual culprits of increased ultra-violet radiation from ozone thinning, acid rain, river pollution from industrial run-off, agricultural herbicides and the destruction of natural habitats.

Another possible cause could be the common platanna *(Xenopus laevis)*. These South African frogs were exported all over the world for use in laboratories and, curiously, in human pregnancy testing. They were recently found to carry a pathogenic fungus which doesn't affect them but can wipe out other frogs. The die-offs have been recorded mostly in those places to which it was exported. The malaise might go down in natural history as Platanna's Revenge.

Burger suddenly got a faraway look, cupped his hands to his ears, then squelched back into the pool. A moment later he was back with a tiny, utterly beautiful frog with a dashing racing stripe down its back.

'Say hi to a micro frog,' he said. 'Not many people have seen it.' Then headed back to place it exactly where he found it.

After our frog-hunting trip I came across a management plan which had been drawn up for Kenilworth Racecourse. It turned out the place was more than merely a refuge for one of the world's rarest frogs. It's an absolute treasure trove. The course was founded in 1882 and the layout of the track has changed little since then. So while roads and houses gobbled up the sand-plain fynbos and horses thundered round the perimeter year after year, the centre was blissfully abandoned to nature.

Some 220 plant species have been listed there, 19 of these endangered Red Data Book species. It's host to some 79 species of bird, including the rare marsh owl, and 13 species of frog. Of the latter, one is vulnerable (the Cape rain frog), one is endangered (the Cape platanna) and the third – the micro frog – is listed as critically endangered.

The area has more than 21 Red Data species a square kilometre, certainly the highest of anywhere in the world. It's one of the hottest spots for endangered creatures on earth. That's one big gwaak for a small racecourse. All power to the Sport of Kings.

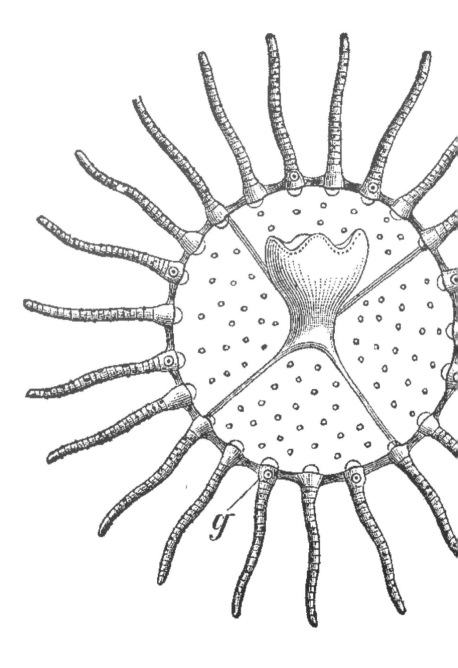

walking things

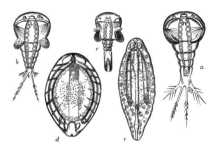

Ships of the desert

It seemed a good idea at the time. Camels are pretty strange creatures and a nose-on photograph would capture that very well. Especially with a close-up lens. Through the viewfinder the camel in question looked oddly like a rabbit. As I moved in for a good shot, its split upper lip curled into what I took to be a rather endearing grin. It wasn't.

What happened next had an effect not dissimilar to tossing a can of smoked mussels into the back end of a fan. I will not treat you to a more detailed description except to say never, ever, allow yourself to be sneezed at by a camel. It's the only time they break their rule of being frugal with body fluids.

As the camel pulled back its head in the haughty pose of its ilk, I cursed it in terms unfamiliar to central Sahara. My only consolation was that the Tauregs I was with hadn't seen the incident. I had a strong suspicion they'd regard a face full of camel snot to be fair comment about those of us who are not from the desert.

As we swayed our way across the peach-pink sands and through the weirdly shaped, sun-blackened rocks of the Akakus Mountains in southern Libya, I became increasingly interested in my ungainly ship of the desert. Despite our earlier encounter it was, I had to admit, quite comfortable – and it left me time to enjoy the awesome beauty of the Sahara.

Arabs, it seems, forgive the camel its arrogance. They say its demeanour stems from the fact that while humans know 99 of the names of Allah, only the camel remembers the hundredth.

But in countries beyond the great sand seas the creature has little to recommend it beyond being a butt for humour. The most positive recorded comments I could find were from a sea captain named Crowninshield, who imported two into America back in 1701 (and who no doubt hoped to make money from them): 'These stupendous Animals are most deserving of the attention of the curious, being the greatest natural curiosity ever exhibited on this continent.'

The eleventh edition of *Encyclopaedia Britannica* obviously took the opposite view. It described the camel 'from first to last an undomesticated and savage animal rendered serviceable by stupidity alone. Neither attachment nor even habit impresses him; never tame, though not wide-awake enough to be exactly wild.'

A book of quotable quotes I have says, most unkindly, that whereas the wheel is one of mankind's cleverest inventions, the camel is one of God's clumsiest. It's a good thing camels are patient to insult.

Given this mention of the wheel, I was rather startled to come across an obscure book on the relationship between camels and wheels by Richard Bulliet, published in 1975 to, I'm sure, a very limited readership. Camels, according to him, are the only form of transport in the history of human technology to have replaced the wheel.

From its first invention to the development of the motorcar, every advance of the wheel has been examined and speculated upon. Whippletree, leaf spring, cambered spoke, each has come in for its share of attention. There has even been speculation as to why relatively advanced societies such as those of pre-Columbian America never made use of this marvellous transportation device.

Yet it appears that, apart from Bulliet's work, there has never been an investigation of why a vast area of the globe, encompassing some of its most advanced societies, chose to abandon its use in favour of the ungainly camel.

But it goes much further. For example, without the camel there may have been no Renaissance in Europe. All this needs a bit of explanation.

During the Eocene period, some 50 million years ago, the camel's ancestors were rabbit-sized mammals in North America. During one of the glacial periods they crossed the Bering Strait and spread east across Central Asia (where they developed two humps) and south into the Sahara (where one hump was somehow more efficient).

In both places this evolving camel developed a defence strategy against predators which was, in biological terms, peaceful and innovative – the ability to live on ice and sand deserts in places few predators could survive.

Camels also have a prodigious capacity to hoard food and water in odd places. Contrary to popular belief, they don't store water in their humps: that's where they keep extra fat. The large quantities of liquid they consume are stashed away in tissues throughout their bodies. They have efficient kidneys which can retain high concentrations of impurities so they don't have to urinate much. They also don't sweat like other animals, instead allowing their blood temperature to rise to levels which would kill most other similar-sized creatures.

To complete the adaptation to flight rather than fight, they evolved large flat feet and (in the absence of shady trees) bulging eye sockets which protected them from fierce, vertical light. A perfect, biological 4x4 waiting for someone to figure out how to use it.

This is where the wheel comes in. The first wheeled culture to penetrate the tribes scattered along the North African seaboard was the Phoenicians. They were followed by the Greeks, who built the magnificent cities of Cyrene and Apollonia, and the Romans, who built the even finer cities of Sabratah, Oea, Leptis Magna and Carthage.

Roads spread, as did the use of chariots. (Within rock overhangs in the central Sahara we were to find ancient paintings of these chariots hitched to speedy-looking horses.) What happened next took more time than can be told in a few sentences, but essentially North Africans, with their backs to the hostile desert and waves of invaders on their shores (including the Vandals), invented a saddle which made it possible to sit on a camel's hump. Before that camels had been hunted and eaten or, at best (for camels), used to carry heavy loads. Now they could be ridden. And for their riders, the age-old fear of the desert evaporated.

The saddle also made possible new weaponry and fighting tactics which shifted the balance of power in the desert. Arab warriors seized control of the caravan trade, and throughout North Africa both riding and pack camels rapidly replaced the wheel. They could carry heavier loads over greater distances than the wagons of the time, keep going for longer and didn't die of heat exhaustion like oxen and horses.

The replacement was to have far-reaching consequences – and pay off handsomely. The whole Arab world soon set its rhythms by the camel: Cairo's streets were made to the width of two packed camels, villages sprang up one day's camel ride apart and great caravans crossed the Sahara all the way to the forests of West Africa. Former nomads and cattle herders became the new lords of the shifting sands.

Which brings me to the Renaissance. Its flowering is generally agreed to have coincided with the waning of the Middle Ages and is considered to have begun in Italy in the fourteenth century. It was a rebirth of culture, art, science, geographical discovery and secular free thought. And it laid the foundations of our modern world.

It's no coincidence that the Renaissance began in Italy. Through its former colonies that country had strong links to North Africa. And by extreme good fortune Arab caravanners discovered great salt pans in the central Sahara for which they found a ready market in hot, saltless West Africa. What they also came across were traders known as Dyula who were prepared to exchange salt for its weight in gold.

Entrepreneurship, it seems, is embedded deep in the human heart. Despite political intrigues, conflict and old animosities, a route was set up from mines deep in the West African forests, through towns such as Kumbi, Gao, Jenne and Timbuktu, across the sands to Tangier, Tunis, Tripoli and Alexandria and on to the goldsmiths and mints of Europe.

The requirements of any good Renaissance is, of course, the wealth to buy the leisure time to pursue all that art, culture and science. And that wealth was not only born northwards on the backs of camels, it was made possible because this arrogant, shambling animal made desert travel possible. It is to the camel in part, therefore, that we owe the Renaissance.

There is, of course, an underlying irony here. The stark truth of it became clear when we happened upon a vast herd of grumbling camels snaking picturesquely across the hot sand.

'Is it a caravan?' I enquired.

'No,' said my guide. 'They're for meat.'

The Renaissance eventually gave us the science to make the sort of wheels that were to win back the desert. Camels – those not destined for theme parks or kept as Taureg runabouts – are now heading for slaughterhouses in great numbers.

So to return to my point of departure: by the time our little caravan had reached the outrageous finger of sun-blackened rock where we'd planned to meet a 4x4 hired to whisk me out of the desert, I had forgiven my camel for its earlier misdemeanour.

A single command had urged it to heave up from its convenient crouch, gentle pressure on its neck with my bare toes had served as an accelerator, and a light tug of its single rein brought it onto its chest so I could hop off onto the sand. The Toyota Land Cruiser, when it appeared, seemed quite crass by comparison.

I bid the Tauregs farewell and watched my camel disappearing into the dust cloud of our churning wheels. Then I turned to pick bits of by then dried snot from my camera. It was a pity, I reflected, that my introduction to this remarkable beast had been by way of its bilge pump. But then I only knew one name of Allah so I could hardly expect to receive its favourable consideration.

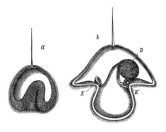

Wild steeds of the desert

It had the makings of a joke, crouching beside a drinking trough in the blistering Namib Desert waiting for a horse. What eluded me was the punch line. After half an hour it wasn't even remotely funny.

On every side the bare gravel plain rippled skywards in shimmering mirages, broken only by the black ramparts of a dormant volcano named Dik Willem. There was no way, on that featureless terrain, that I could approach the Namib's skittish wild horses. The only sensible plan had been to wait for them to become thirsty enough to head for the water piped in by park officials. Some binocular work had suggested dots on the horizon were, indeed, horses and that they seemed to be heading my way. So I hunkered down, camera in hand, feeling like a loaf in an oven. And waited.

In the deep silence the soft plop of a hoof was clearly audible.

I didn't dare peer over the low stone wall separating the drinking troughs for fear of startling the scout stallion. He suddenly loomed round the wall, huge, chestnut and decidedly wild-looking. Seeing me, he threw his head up and snorted. I didn't move and looked down, trying to appear non-threatening. Then he backed off behind the wall and I heard slurping.

Soon the rest of the group arrived, each eyeing me owlishly but not taking fright. I slowly raised my camera. At the soft click of the shutter every horse in sight jerked its head up and seemed ready to take off: in the silence of the great desert small sounds clearly carried great portent. Maybe it reminded them of a rifle bolt sliding into place. But my second shot raised less interest, and soon they tolerated my camera's attention.

They were large, long-limbed beasts, these survivors in a place where logic suggests no horse should be; thin but otherwise in good condition. There was evidence of some nasty battles, however: hacked ears, kick marks on their sides and bites on their withers. Their hooves, also, were chipped from the harsh desert gravels.

For many years the presence of these feral horses was not widely known, their range falling within the Sperrgebiet – the forbidden zone in which diamonds are mined. Where they come from is a mystery with many theories, though most people agree they have been there since 'German times' (before 1919).

One theory has to do with a cargo steamer loaded with thoroughbreds which escaped when it ran aground near Lüderitz. Another that they're descendants of Nama horses brought into the area at the beginning of the twentieth century.

A rather more romantic tale concerns an eccentric German nobleman, Baron Hansheinrich von Wolf, who built a castle named Duwisib on the edge of the desert. He imported some fine horses from Europe and bred stallions for the German military Schutztruppe patrolling the Sperrgebiet and protecting natural springs near Garub and Aus. When war was declared in 1914, the baron joined up and was killed in Europe, leaving some 300 horses round the castle to their own fate.

Early in the same war South African forces cornered the Schutztruppe and interred them at Aus. In the confusion of the trench-based battle, the German steeds may have broken free and gathered at the natural springs which they'd remembered from patrols.

Horses are hardy creatures and, since the Lower Eocene Period some 55 million years ago, have wandered the plains of Europe, Asia, Africa and, later, America. But for centuries – until the invention of the internal combustion engine – they were too useful to be left to wander unharnessed. The last truly wild horses were *Equus przewalski*, which thundered across the icy Mongolian plains a hundred years ago. *Equus caballus* – the standard issue horse – has been fully domesticated for maybe 5,000 years. But give it a gap and a domestic mount will take to living wild like a duck to water. All it needs is grazing and water. In the Namib Desert, on the face of it, these things are in short supply.

The area in which the horses live consists of gravel slope, dune sand, tough calcium carbonate-permeated crust and a hard stony layer known as desert pavement. In a good year perhaps 40 millimetres of rain falls. But it doesn't hang around. An area is considered to be a desert if evaporation is twice the precipitation rate – in the Namib's vast dune fields evaporation is around 200 times greater. Still, some grass does grow.

For many years the Namib's feral horses survived on natural springs, supplementing water intake from the Garub rail siding where a borehole was dug in 1908 to serve the now-defunct steam trains. When the springs dried up and the station closed down, De Beers Diamond Mines installed a drinking trough for the horses. It's now their sole water supply, but today is maintained by officials of the Namib-Naukluft Park.

A year-long study of the horses done by Telané Greyling of Potchefstroom University shows just how adaptable they can be. She found their diet included a range of tough grasses as well as some desert shrubs and twigs – all of which added up to around seven kilograms of nosh per horse a day, about the same as any pampered pony.

It's just that it takes a whole lot longer and far harder work to get what they need. So they sleep less: around four hours a day instead of a normal horse's seven. Hardly surprisingly, they spend most of their time grazing: far longer than their greenfield cousins. They also play less (even the foals), walk slower to conserve energy and have little time for comfort activities such as rubbing, rolling or even yawning. The stallions posture more and fight less than domestic horses, though when they *do* fight it's all gnashing teeth, flying hooves and dust.

Much of their energy is taken up trekking long distances between grazing and the water point, so the horses have found a solution. They can easily go for up to 72 hours without a drink (the longest time recorded by Greyling was four days).

They also eat dry dung (it's named coprophagy) for the simple reason that, despite having passed through a horse, it still has a higher and more concentrated energy content than the surrounding grass.

The horses I came across looked in fine shape, but the desert takes its toll. Greyling found that an average of around four in ten foals died not long after birth. The causes included predation by jackals, hyenas or possibly even leopards, nutritional stress, getting left behind because of worn-down hooves, and just plain exhaustion from walking too far. Occasionally one wandered onto the almost dead-straight road between Aus and Lüderitz and was hit by a vehicle.

There is, of course, a question that needs asking about the feral horses of the Namib. Should they be there at all? In 1986 the diamond area – which includes their 40,000-hectare range – was taken over by the Namib-Naukluft Park. Horses are not part of the native wildlife and must surely be competing with oryx, springbok and steenbok in the area.

There are those who argue that the horses are of historical and scientific value and a great tourist attraction. Others say they should be left to fight for their own survival without human intervention. Conservation purists would have them removed. If they *were* trucked out and domesticated, the 150-or-so horses would probably soon die, having developed no natural immunity to viral infections.

Those who would like to see the horses remain in the Namib *could* claim that the creatures are merely returning to an area in which their ancestors once ran wild. The problem is the nature of the ancestors, the notion of an 'area' and the timing of 'once'.

Palaeo-anthropologists agree that three-toed, horse-like creatures known as hipparians would probably have left their bones beneath the desert sands. But that would have been maybe five million or more years ago. So, genetically speaking, the present steeds have a very slender claim on grounds of succession. Anyway, in those far-off days the Namib could well have been tropical jungle. So, while it's true that if humans hadn't exhausted the natural springs the Namib horses would have been able to survive unaided through physical and behavioural adaptation, this doesn't say their tenancy is environmentally balanced.

A better argument is the historical one. For around 80 years, with no murmur from a conservationist, the Namib horses have adapted to their environment with only a little help from their friends. For these refugees from human misadventure and neglect it's been a heroic struggle. In the process – if you'll excuse the expression – they've certainly earned their desert spurs, not to mention a right to our respect.

And there's something else. Unlike a cow, sheep or dog – and rather like a cat – *Equus caballus* has never been fully domesticated. There is about it something of the flightiness of Pegasus, the raw power of the Centaur, the ethereal beauty of the unicorn and the wildness of Poseidon's wave steeds. But these are all mythical. The wild horses of the great Namib Desert are real.

I'd been crouching beside the water trough for long enough to develop a painful cramp in one leg. When I judged the horses had drunk enough I stood up – which resulted in a thunder of hooves as they galloped back into the shimmering desert, long manes and tails streaming.

There are few sights more stirring than wild steeds flying across a fence-free plain. If the only place they can do that is in an unwanted desert, it seemed to me, then let it be.

Monkey business

Gorillas are simply outrageous. Nothing prepares you for meeting one on the green-dripping, moss-covered, butterflied equatorial forest floor. They look up at you from their wrinkled, black leather faces and it's ... it's ... well, it's extremely difficult writing about mountain gorillas. Words seem woefully unable to convey the emotional impact of the experience. When one first locks onto your gaze with its beautiful, wise, hazel-brown eyes your ears ring. It's a sort of First Contact; it whacks you in some ancient corner of your animal brain and comes out as tears. When the gorilla looks away you feel instantly lonely.

I didn't know that about gorillas until I met one, of course. What I was thinking about when I boarded Air Uganda's only plane in Johannesburg and headed for Entebbe was people. I'm not sure when it first occurred to me that human beings might be an evolutionary mistake: probably while watching the eight o'clock news. Sure, we've taken over the planet, but judgement about the genetic path we're on really depends on whether you rate success as the ability to loot, burn and pillage or live in harmony with earth's other life forms. If we are on the wrong track, I got to thinking, where and when did we branch off?

There's heated debate in some scientific circles about whether we first stepped onto the savanna and stood up because the forests receded and the grass was tall, or became a semi-aquatic, hairless, dolphin-like creature able to hold our breath because the forests flooded and stranded us on soggy islands. But, either way, we probably began the stooping march to cellphones and hamburgers in the equatorial forests around the Great Rift Valley.

We left them, conquered space and invented paper clips. But gorillas and chimpanzees stayed put, almost unnoticed by the human world until fairly recently. With logging operations and banana *shambas* hacking away at their ancient forest homes, however, these distant cousins of ours are now under terrible threat.

Before leaving South Africa I was not sure if looking into their eyes would count for much, but I wanted to visit them in the wild before we turned their habitat into a coffee plantation – to somehow say sorry and to see if, maybe, it was they (and not we) that had taken the more sagacious road.

My introduction to Uganda, however, was not primates but lake flies. Around the Great Lakes they're not measured in billions but in kilotons, and form the base of those food-chain pyramids you see in biology books with humans perched at the top. From the aircraft they look like brown mist floating over the water and their swarms are so dense people have died in their midst, unable to breathe. When you swallow them by mistake they won't go down, but seem to stick in the back of your throat.

Entebbe is notable for its international airport, the gracious

Windsor Lake Victoria Hotel and monstrous Nile perch which get dragged out by locals (who think a 30-kilogram fish is not worth mentioning at the sailing club). From there we joined a reasonable tar road carrying a veritable river of taxis, little Japanese motorcycles and hurtling buses towards the equator and (for those who persevered) the Congo border to the west. First stop was not gorilla-type jungle but Queen Elizabeth Park to see wild chimpanzees, the sort that don't ride round on unicycles wearing sequined fezzes.

Mweya Lodge is a magnificent place overlooking both the Kazinga Channel and Lake Edward. Hippos and warthogs keep the grass short and a nosy family of banded mongooses patrol the chalets searching for rhino beetles and lesser insects, which they crunch, somewhat disgustingly, under your table or chair. But its value, for me, was its proximity to a secretive primate haven.

Chambura Gorge sneaks up on you and its appearance – a deep forested gash across the Rift Valley floor – can take your breath away. Its name means 'search and fail', an appellative it earned because of the many local people who entered it never to return. As I peered over its rim onto the top of the gallery forest below, a violet-blue Ross's turaco, perched atop a towering Uganda ironwood, 'kkkowed' in fright and took to the air, flashing brilliant red underwings. Black and white colobus monkeys, looking like little bearded men in dress shirts and tailcoats, squinted up at us comically from the top of the canopy, then went on foraging.

A steep path led into the gorge and, as guide Tushabe Venantius shepherded us down, a red-eyed dove took up its usual complaint: 'Oh dear, my eyes are red. Oh dear … ' Like all other guides I spent time with in Uganda, Tushabe was absolutely first class – committed, knowledgeable and easily able to identify animals, birds or plants in both English and Latin. 'Take photographs,' he told us with a wide grin, 'but leave only footprints.'

Down at river level a troop of olive baboons crashed across the path, trotted along some horizontal boughs and lowered themselves into the abundant undergrowth. Soon afterwards we heard the chimps.

'They're eating *sabu*,' chuckled Tushabe. 'It's a fruit – a bit alcoholic – and it makes them talk a lot.'

The hard, tennis-ball-sized fruit thudded down almost at my feet before I realised we were beneath the chimps. For a moment I couldn't make out what my binoculars had focused on as I swung them upwards, then realised I was looking at the hairy backside of a large primate perched comfortably above me scoffing bright yellow fruit.

The chimp leaned forward and peered down, looking slightly peeved, took a bite then peered again, as though he had second thoughts about the creature gaping up at him. It could have been my imagination, but his expressions seemed both human and entirely understandable.

His next action was so elegant that if I wasn't glued to my binoculars I'd have applauded. He stood up, with one hand grabbed the branch he was standing on, swung under it (still holding with one hand), let go at precisely the right point in his swing to catapult himself, spread-eagled, onto a lower cluster of leaves way too thin to hold his weight. But he merely held on as the branch bent, then let go as his feet were deposited neatly on a lower bough. Then he sat down, gave me a hard look and peed loudly onto the leaf-covered forest floor. His last action left me in no doubt who the alpha male was around that neck of the woods.

As we left the gorge, a large male lion broke the cover of a euphorbia thicket, bounded across some open grassland and dived down a path into the gorge we'd just vacated. Somehow I hadn't reckoned on lions, but it made me remember the place's name: Chambura, where people go but do not come back.

From Queen Elizabeth Park the tar headed east, but we soon turned south into a maze of un-signposted tracks which had more in common with river beds than roads. I can't imagine how anybody not born and raised in the area could make sense of them. Side roads constantly veered off, particularly just before blind bends, and it took a while to work out that they were part of an ingenious system of splitting roads round curves to prevent accidents. An unwary traveller taking the wrong one by mistake would be almost assured of a head-on collision with one of the careening vehicles which hurtled past from time to time.

'Godammit!' yelled tour guide John Addison as the umpteenth oncoming minibus refused to vacate its mid-road position. 'They think they bought the road with their licence!'

A bakkie, with what seemed to be an entire village on the back, roared up behind us, hooting urgently. There was no room to pass but the hooting continued until it managed to ram past on the obligatory blind rise. We looked in wonder at the massively overloaded vehicle – a moving tribute to both Toyota and mango-tree mechanics.

The villages we passed were mostly neat rows of mud houses amid seemingly endless banana fields. Many doubled as shops and sported signs such as Another Life Saloon, New Vision Off-Sales, Set-Set Hair Salon, Tender Teapot Hotel and, along a particularly nasty stretch of road, Doctor of Broken Bones.

At one small town we needed to make a pit stop and pulled in at a neat little hotel and went in search of a loo. It was a pit all right, a squat-and-swat which seemed to have been positioned so locals lining the veranda could watch *mzungus* staggering out with expressions of shock on their faces. 'There are probably so many diseases down there', commented one of our party in his broad Scottish accent, 'they'd scare the mould out of penicillin!'

After hours of chassis-punishing lurching and banging southwards towards the northern border of Rwanda, the scenery suddenly rose up ahead of us, impossibly green, and we turned down a side road (I use the term loosely) marked by an alluring sign: Bwindi Impenetrable Forest.

At Mantana tented camp we were greeted with a tray of iced lemon drinks and another of cool, rolled face cloths with which to smear off the dust. Birdsong rose from the forest and the smell of cooking drifted across from the kitchen. From the camp we could see the canopy of the brooding forest, threaded through with wraiths of mist. A tropical storm rumbled ominously in the mountains beyond and the damp, warm air felt like the breath of a living creature. It must have taken an awful cataclysm to force our early ancestors out of such a paradise. From somewhere a phrase was downloaded into my primeval memory: here be gorillas.

There are three sub-species of gorilla: western lowland, eastern lowland and mountain gorilla. The third – *Gorilla gorilla beringei* – is by far the rarest, with an estimated 600 in existence, and is found only in the high, Afro-mountain rainforests in the vicinity of the remote Virunga volcanoes in Central Africa. They were 'discovered' when two were shot by a German army officer, Oscar von Beringe, on the slopes of Mount Sabinyo in 1902. Now, ironically, the sub-species bears the name of their assassin.

Their habitat, overlapping Uganda, Rwanda and the Congo Republic, is politically volatile: until recently it was a battle zone, with thousands of refugees and soldiers trampling through the forests, exposing gorillas to gunfire and human diseases (97.7 per cent of gorilla DNA is 'human', so they're susceptible to most of our ills).

Their lowland cousins, though more numerous, are increasingly falling prey to effects of mainly European-based logging companies which cut roads into the virgin forest, and from hunters who use the roads for access into their habitats. 'Bush meat' is the main source of protein for people in the region (and for loggers) and it is estimated that some 40,000 tons of it are consumed each year in the Congo alone. Primates are part of this plunder, and around 600 gorillas and 3,000 chimps a year end up in cooking pots. Given their genetic proximity to humans (chimp DNA has a 98.6 per cent human overlap) this virtually amounts to cannibalism. It's like eating your ancestors.

Situated in now-peaceful Uganda, however, Bwindi Impenetrable Forest is a safe haven. There, in relative security, the great, lazy primates wander, rest and sunbathe between bouts of eating and sleeping. Apart from the occasional luxury of an ant *hors d'oeuvre*, gorillas are gentle vegetarians, nibbling the leaves and stripping the bark from around 58 plant species, then belching luxuriously as they rest their bloated stomachs in supine majesty.

After formalities with permits and the selection of trackers, we met with our guide, Richard Magezi, at the entrance to the park. In 1991 he began the two-year task of habituating the Mubare group to human presence, and now considers them virtually part of his own family. He outlined the rules: no more than six people on the trek; nobody with any illness permitted near gorillas; approach no closer than five metres; and maximum contact time one hour.

As we entered the forest a red-tailed monkey dropped from a boscia branch *(Boscia coriacea)* in the canopy, its tail streaming out behind like the cord of a bungee jumper. A chimp – dissecting nuts along another branch amid a flap of great blue turacos – took no notice.

A squelch of earth in a rare shaft of sunlight had attracted a crazy whirlwind of butterflies. Gaudy swallowtails, blue mother of pearls and chocolate browns dominated a mêlée of smaller white, orange, red and speckled flutterers, all competing for places to slurp the ooze with their outrageously long tongues.

The previous day the Mubare group had been spotted in the next valley and we made for that point, following the machete-swinging trackers through impossible-looking tangles of branches, leaves, ferns and wicked stinging nettles. At times we were moving on packed foliage up to a metre above the forest floor.

When we arrived at the place where the gorillas had rested the previous day, I picked a half-eaten leaf and a chewed stick and tucked them into my daypack. Somehow it seemed significant to keep the leftovers of a gorilla lunch. From there the tracking began in earnest and I soon discovered the benefits of walking on your knuckles: where gorillas had passed with ease, we humans slashed and cursed, got caught by vines and were smacked by overhanging boughs. In the jungle, bipedalism was bad news!

'Shh! The gorillas are here!' whispered Richard suddenly and everyone froze. I detected the movement of a dark shape ahead and stared fixedly at it. Then, glancing to my left for no particular reason, I found myself in the gentle gaze of the most thoughtful brown eyes I'd ever seen. The female gorilla was sitting like a silent, furry Buddha only a few paces from me, exuding a peacefulness which offset any possible fear I might have had in the presence of such a powerful, near-mythical creature. Then she tipped onto her knuckles and loped to the base of a giant ebony tree *(Diospyros abyssinica)*, lay on her side and began fishing for termites, licking them off her fingers and grimacing comically when they bit her.

We moved a few paces and were halted by the presence of an enormous silverback. I remembered Richard's instructions if he charged: crouch down and don't make eye contact. But I could not drag my eyes away from him.

Beneath his huge crown were two penetrating eyes, a shiny black leather face, enormous air-scoop nostrils and a mouth you'd have to describe as quizzical. His muscular arms reminded me of Popeye and his torso would be the envy of a sumo wrestler, but my startled gaze was drawn to his fingers: they were the size of huge, tropical bananas. Heaven help anything that fetched a clout from a silverback! He rumbled deep in his throat, causing me to fear the worst, but then ambled off, with us skulking in his magisterial wake.

'Come quickly,' hissed Richard after a few minutes of trying to tiptoe through underbrush with the consistency of newly boiled spaghetti. We peered round a bush and there the great creature was, comfortably scratching his broad buttocks with an expression of complete contentment. Beyond him were three females, another young male, some adolescents and two babies.

A youngster – looking for all the world like a cuddly toy – bounded towards the scratching patriarch, sat down beside him and pounded his little chest, then looked up at dad for approval. Having secured that, he leapt for a branch, hung by his feet with his arms dangling and offered us an upside-down grin.

The silverback glanced at his gawking audience with not a trace of interest – we could have been forest butterflies for all he cared – then rolled onto his giant knuckles and was gone.

Our paths had parted. But where they would ultimately lead remained an unanswered question. By the time we'd bone-jarred our way back to the howling madness of downtown Kampala, however, I had no doubts about which branch of the family tree I'd rather hang out with.

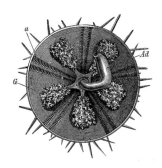

Lords of the trees

The love affair began in the Impenetrable Forest. It was a commanding performance: with an ear-numbing crack, lightning slammed into a giant mahogany towering above the rainforest canopy, unleashing a tropical deluge which bent the giant trees, sending leaves spinning to the ground. Huge, tightly packed drops seemed to squeeze the breath out of the sweaty air and left me drenched, gasping and enraptured.

Half an hour later it was gone, trailing wraiths of sun-flared mist which billowed from the dripping canopy. A black-and-white colobus monkey began its frog-like *gwaar*, *gwaar*, *gwir*, *gwir* call and a family of chimps greeted the sun with roars and pant-hoots.

In local language the forest is named Bwindi – which means impenetrable – and it's situated in southern Uganda on the Rwandan border. My reason for being there was mountain gorillas, monstrous, amiable vegetarians which burp and rumble in extreme rarity along the forested hills and volcanoes which mark the border between the two countries.

The trip had been appealing: to find as many of Uganda's primates as possible. Along for the adventure were primatologist Mike Lawes of Natal (Pietermaritzburg) University, guide Medi Lwere, and John Addison of Wild Frontiers. There are some 17 species of bushbabies, monkeys and apes in Uganda – outside the war-ravaged jungles of Congo, it's the best place in Central Africa to find primates.

A mere 15 minutes from the camp we came upon the gorillas in low, secondary forest; a huge silverback and his family contentedly munching leaves and wild asparagus. They ignored us completely but for a youngster, who responded to his gawping audience by hanging upside down to see whether we looked any better that way round.

That night we saw a potto, a strange, nocturnal, tree-creeping prosimian whose huge eyes gleamed eerily from his high perch. Wrapped in the silent darkness of the forest floor, watching our probing torch beams lance the canopy, the monkey in me began to awake: I had, somehow, been there before.

Some 60 million years ago pottos and bushbabies – prosimians – split from our family tree, and around 30 million years later another divide occurred – guenons, baboons and the colobus family taking one branch; gorillas, chimps and humans taking the other.

Chimps and humans are almost the same creature, sharing more than 98 per cent of their genetic make-up. There's an interesting theory about why we look so different. Some reputable scientists, such as South African palaeontologist Philip Tobias, speculate that the two per cent difference may have occurred after our ancestors took to the water, becoming aquatic apes, standing upright to keep their heads above water and also freeing their hands. We are, after all, the only 'naked ape', using a fatty layer under the skin rather than hair to keep warm – just like a dolphin.

Well, maybe ...

Our trip had begun in Entebbe several days earlier and we'd arrived at Bwindi by way of Mgahinga, a park which clings to the slopes of six volcanoes in the Virunga Mountains. The renowned primatologist Diane Fossey had frequented the little hotel in which we slept. Our hope had been to find golden monkeys, but they'd stayed well hidden.

We headed north from Bwindi, and by the time we reached Kyambura Gorge – a forested gash in the rippling savanna just east of Queen Elizabeth Park near Lake Edward – we'd added black-and-white colobus to our list. But the chimpanzees we'd hoped to see were absent. Arboreal primates, it seemed, were damnably difficult to find.

From Queen Elizabeth National Park we bumped our way up the eastern edge of the Ruwenzori Mountains – the fabled Mountains of the Moon – to the busy little town of Fort Portal. The next day we entered Kibale forest.

There was something so uncanny about the place it gave me goose bumps. The rainforest was ancient and enfolding, huge yet accommodating: it was like walking into a great medieval castle and realising it felt like home.

Ebony, mahogany, teak and countless other trees, buttressed on wide, fluted root systems, soared upwards through the canopy some 40 metres overhead. Brightly coloured bracket fungi step-laddered up dark, rotting stumps, lurid mosses lurked in gloomy wooden root crotches and – in shockingly bright shafts of sunlight – butterflies of many species pirouetted and played. The whole space seemed to be bound together by lianas and threaded with birdsong.

Kibale consists of more than 750 square kilometres of rainforest, swamp, grassland and bush – and one of the highest density of primate species in Africa, including around 600 chimps. We soon found them – atop a huge fig tree *(Ficus mucuso)* nesting, copulating, peeing and stuffing themselves on fruit. Occasionally a barking argument broke out, and the forest rang with their whoops and yells.

Staring up at their hairy bottoms, it occurred to me that, in a way the Big Five embody the spirit of the African savanna, primates bring the rainforests into focus. They are the laughing, squabbling, leaping lords of this domain. Without them the great equatorial forests would be lifeless. Without the forests they would be caged, sad monkeys – or pot roast.

That afternoon I trotted off alone to the forest edge and wandered into its green enfoldment. It was raining lightly and the canopy leaves, which all have clever drip points, tipped little streams of water down my neck. To avoid a soaking I curled up between the root arms of a huge mahogany and – for no reason I can imagine – went to sleep.

When I awoke the rain had stopped. An olive baboon was peering at me nervously through the undergrowth and, nearby, grey-cheeked mangabeys were having an intense 'whoop gobble gobble' conversation. As I stood up the baboon blinked rapidly and vanished without a sound. I picked my way back to the forest edge in a daze somewhere between sleep and bliss.

Along the way I remembered something the ecologist Norman Myers had written about being in a rainforest and I later dug it out: 'Something in the forest around me served to stretch something inside me ... a host of additional nerve endings came alive. I felt my whole being was standing on tiptoe.'

The pain would come later, in a university library surrounded by books on rainforest ecology: the rainforests are dying, and most people don't seem to care.

Those great tree masses which girdle the planet's tropics are astounding reservoirs of biological diversity. A square kilometre of rainforest has the same weight of trees as 200 to 300 square kilometres of woodland. One study found 138 tree species in a single hectare. In Africa roughly half of all bird species are dependent on rainforests.

They are among the oldest, most stable and most complex ecosystems on earth. These natural ancient monuments have outlived most early types of vegetation. Many of the trees themselves are living fossils, exhibiting forms believed to be characteristic of primeval trees.

Fossil evidence suggests these forests have been unchanged and undisturbed for around 100 million years. In the ice ages they endured, providing vital ecological refuges for creatures escaping the freezing wastelands – and thereby becoming hothouses for new species.

Their presence regulates weather, rainfall and the amount of carbon in the atmosphere. Despite this, humans are cutting down – according to the United Nation's Food and Agricultural Organisation – 168,000 square kilometres of rainforest a year. That's more than 1.5 million square kilometres – an area larger than South Africa – each decade. With the forests go the creatures within them ...

Although these rainforests cover only about one-fifteenth of the planet's surface, they are thought to contain anything from 50 to 70 per cent of its species.

A rough estimate suggests that, at current deforestation rates, the world will lose between two and seven per cent of its species in the next 20 years – that's between 8,000 and 28,000 species a year, or 10 to 75 a day. Species are disappearing before we even know they exist. Unless some drastic curbs are put in place, by 2040 more than one in three tropical species will have been committed to extinction.

Bye-bye, lords of the forest. I guess our ten-million-year relationship couldn't last forever.

From Kibale we drove up to Murchison Falls National Park on the White Nile – a rough trip to a beautiful place – then back to Entebbe for a boat ride to Ngamba Island Chimpanzee Sanctuary in Lake Victoria.

The island is a square kilometre of high forest and was bought by five concerned international organisations as a chimp sanctuary. Most of the chimps there have been saved from hunters who killed their parents for bush meat. Goodall found that for every live chimp taken from the forest, at least six have been killed. Others on the island have been liberated from zoos or some other form of captivity.

When youngsters arrive they're generally traumatised and need constant love and attention. This is done by a dedicated group of 'chimp mums' who carry them round, feed them and tuck them up at night.

I was sitting, staring at the island's patch of rainforest, when chimp mum Heather Cohen walked past. Her charge, Pasa, chittered and held out its arms in my direction, so Heather plopped her on my lap. The chimp snuggled into my jacket and held my hand with its strong, leathern fingers.

Chimps are extraordinarily smart; you just have to look into their eyes to know it. Sue Savage-Rumbaugh, who spent many years teaching and studying a pigmy chimp named Kanzi, said that working with him was like 'standing at the precipice of the human soul, peering deep into some distant part of myself'. Gazing at the little chimp on my lap, I knew what she meant.

We sat there for a while, me feeling really down about the bushmeat trade and the chimp feeling who-knows-what. I wanted to say sorry to Pasa for the wanton looting of her ancient home and the destruction of her family. But I couldn't speak Chimp.

The little ape stared at me intently for a bit, looking a bit puzzled, then pursed her lips into a cute oval and emitted a series of soft, rather sad hoo, hoo's. In the circumstances it was precisely the right thing to say. Then she squirmed round, leaned forward and planted a hard, reassuring kiss on my lips.

The long, dark road of the cave cricket

Grunthos the Flatulent dived into the hole in the ground, followed closely by his swarthy cave-goblin master, Peter Swart. I followed Peter's disappearing boots and his dog's vapour trail, leopard-crawling with my backpack in my teeth, the only way I could figure to transport it under the circumstances.

After too many metres of head-banging gloom and a lot of gritty sand in my boot tops, the cave opened out. It was so dark my eyes started to manufacture fireflies in desperation, the torches humbled by the immensity of their task. The gloomy sandstone walls acted as light sponges, and everything not actually in the torchlight circles remained inky black. The only sound was the monotonous dripping of water from sodden rock surfaces.

Our quarry was, by more normal standards, wacky: the long-legged, golden-haired cave cricket, known to science as

Spelaeiacris tabulae. Maybe 50,000 years ago its ancestors and a few close cousins took refuge in caves and never again came out, a fairly drastic move.

The problem for cave crickets began with the annoying predisposition ice has for stealing water. Earth has a gigantic freezer at either end and these – at times – have had a tendency to ground airborne water vapour and turn it into glaciers.

Before this series of climatic hiccups – for millions of years at a stretch – the planet chugged along in comfortably warm, wet conditions, keeping its polar caps firmly under control. In climatic terms, the Tertiary Period – from 65 million to 2 million years ago – was a doddle. The bossy dinosaurs were out of the way – who knows what wiped them out – and other reptiles, mammals, birds and insects had the place all to themselves.

It's not certain just why the weather went wobbly during the Pleistocene epoch. Was it sunspots, meteor-impact dust shutting out the sun, or something else? The result was two million years of ice ages flip-flopping with warm, wet periods. Glaciers swallowed water in great icy gulps, causing deserts, then spat it out again, flooding beleaguered landmasses.

It was a tough time. North America lost its horses, camels, bears, sabre-toothed cats, mastodons and mammoths. Australia is thought to have lost 28 genera and 55 species, including clawfooted kangaroos and *Genyornis*, the biggest bird ever to stalk that land. Africa hung on to more species than most continents but the attrition rate was still high, aided by humans who hunted and tended to set the veld on fire.

The problem cave crickets had was not humans or carnivorous kangaroos but humidity. Thousands of years ago, when the forests of Southern Africa were high and the rainfall torrential, crickets and countless other little creatures lived under moist leaf litter and thrived. It was dark, warm and soggy. When the polar caps expanded and water was scarce, the forests withered and more small species than we will ever know vanished. Some

– very few – sought out dark, warm, soggy cracks and caves and began evolving into troglodytes, surviving in these so-called refugia.

The cave we were in was named, rather unsettlingly I thought, Boomslang, and is part of a system of caverns above Kalk Bay on the Cape Peninsula. I'd trailed up via Hungry Harry's Halfway Halt (don't ask) behind the bearded cave-hound Grunthos and his equally bewiskered owner, Peter, a Cape caver of distinction.

There are about 80 caves in the mountains above Kalk Bay and perhaps ten on Table Mountain, together forming the world's fourth-largest sandstone cave system. They were created by water erosion eating away at cracks, unlike limestone caves. Perhaps fortunately, their entrances are small and often overgrown, which has ensured a degree of protection from human intrusion.

Soon after we'd found a place in the cave where we could stand up, Peter found a cricket. With its long legs it looked more like a ghostly spider, and its spectacularly long antennae waved round in what seemed like alarm as our torches bore down upon it. Light was clearly not high on the creature's list of good experiences. Waving round in the crack like that, the small, lonely cricket had the look of something that deserved to be labelled 'critically endangered' in the Red Data Book.

Early studies suggested these crickets survived on a diet of soggy cave lichens, but Mike Picker and Norma Sharratt of the University of Cape Town, who studied the crickets and other Golem-like creatures in the caves, found them to be omnivorous scavengers. One study found cave cricket faeces to contain moth scales, green vegetables, bat hairs, rodent hairs and human skin. The last item might have been from the body of some explorer who never found his way out, but was more likely from the scraped knees, elbows and craniums of cavers. Feeding crickets could be detrimental to your health.

Swapping forest leaf mould for caves those thousands of years ago required some adjustments. In the inky blackness, eyes became redundant. Everything in there feels its way around, which, given the presence of two other cave predators, provided me with an image of unimaginable horror.

There is a spectacular, blind, snow-white velvet worm in the Cape caves, thought to be a sort of 'missing link' between worms and insects. It has the antenna and claws of an insect, but the body of a worm, and waits, silent and hungry, until a meandering cricket touches it with an exploratory feeler. Then it blasts the unfortunate cricket with sticky slime from a gland, hopelessly entangling it. If that were not enough, the caves also contain large predaceous spiders and ravenous centipedes which specialise in cricket snaring. A cave cricket's life is clearly not an easy one.

Prodding around in Boomslang Cave turned up some other denizens of the dark. Zoology attaches wonderful names to these things. The science of cave beasties is termed biospeleology and the study of their shapes and types is troglomorphy. Then there are humicoles which scoff leaf humus, endogens which live in the soggy soil, and (my favourite) guanophiles which live on bat scat. It's definitely goblin language.

Sharratt and Picker – patient crack- and puddle-sifters that they are – have turned up an astounding 85 species in the Peninsula's caves – including such things as pseudoscorpions, cave earwigs and snow-white cave shrimps – and a clutch of things never before described by science. Many are what are known as relict species, little pockets of refugees from the Pleistocene ice ages seeking climatic asylum.

And little pockets they most certainly are. A species is listed in the Red Data Book as critically endangered when it is found in an area of less than 100 square kilometres. Some of the Cape cave creatures are only known within less than ten square kilometres. Some have been found so seldom they are listed simply as 'data deficient'.

What's really interesting, though, is where their cousins live. The nearest relation to the white cave shrimp is not in Africa but in Brazil and Western Australia. The only similar species to the South African water shrimp are in South America, Sri Lanka and Australia. The harvestman has its nearest relatives in Chile, Sri Lanka, Australia and New Zealand, and cave crickets have family connections in Patagonia, the Falklands, New Zealand and Australia but nowhere in Africa.

Just think for a moment: these are all creatures that cannot live outside of caves and in sunlight. The only way they could have spread across the planet is if the caves they lived in were connected. The idea of intercontinental caves boggles the mind until you realise they were probably all once connected in the supercontinent Gondwanaland. The caves were separated when this massive landmass broke up around 145 million years ago. So the diminutive cave cricket and all its pals and predators are living proof of continental drift. A weighty distinction for such small and virtually unknown creatures.

After an hour in Boomslang I was starting to have serious mind drift. In the absence of antennae and thousands of years of adaptive history, the utter blackness of the caves was becoming seriously disorienting. Grunthos the Flatulent, with an uncanny ability to navigate in the dark, headed off, followed by Peter. The cave grew quite spectacularly, shrank for a while, then began leaking light. We emerged above Fish Hoek, having gone right through the mountain. Sunlight may be anathema to cave crickets but, as its warmth re-heated my bones, I was glad right then that my species took a different adaptive road through the ravages of the Pleistocene.

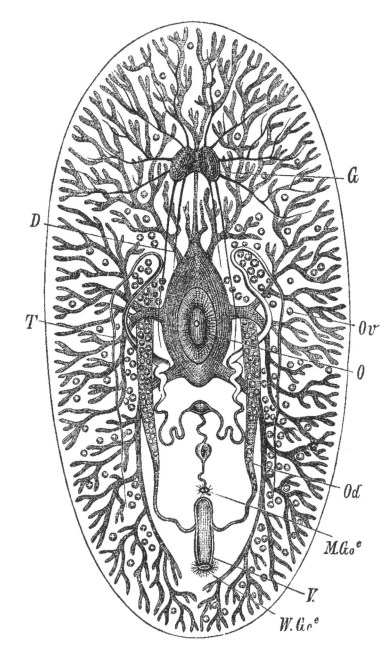

homo sapiens
of distinction

Worker in a science yet unborn

Many considered him to be mad, and there were times when he would have agreed with them. There were other descriptions: muckraking journalist, hypnotist, drug addict, lawyer, surgeon-general, genius, a founding father of Afrikaans literature and one of the finest naturalists South Africa has ever produced. But for Edna Cross, a ten-year-old who shared a house with him in Pretoria, Eugène Nielen Marais was simply Uncle Eugène, the most magical adult she ever knew.

'There is nobody like him today and there will never be anyone like him again,' she was to say thirty years after his untimely death. 'His room was a wonderland. I remember the numerous books, and the skulls on top of the wardrobe – a human skull and an ape skull – and the collection of knives. His stories absolutely enthralled us. A single story could last for weeks ... '

In the summer of 1936 Marais put the barrels of a 12-bore

shotgun in his mouth and pulled the trigger. More than six decades later his stories still enthral us and challenge science – he was that far ahead of his time.

We will never know whether his addiction to morphine enhanced his scientific insights or blocked even greater achievements. The drug was possibly a mistake: when Marais was a baby no well-equipped household medicine cabinet was without an opium-based anodyne for pain, nervous attacks or hysteria. Babies were lulled with concoctions such as Dalby's Carminative, McMunn's Elixir, Mother Bailey's Quieting Syrup, leaving countless youngsters calm but addicted. When, as a young journalist, Marais hit stressful times, it was to morphine that he turned. When his young and beautiful wife, Aletta, died in childbirth it became his companion.

Periods of heavy addiction, and other times of withdrawal, were to turn him increasingly into a recluse – seeking out the healing power of South Africa's wildest places, far from humans who would pass judgement on his deplorable condition. Disillusionment with the outcome of the Anglo-Boer War simply speeded his trajectory. Oddly, science was to gain from this.

Marais was born in 1871 in Pretoria, then still a frontier town. Lions roamed where the Union Buildings now stand. The area round his home teemed with mambas and puffadders and, even as a youngster, he kept scorpions – much to the alarm of his six prudish older sisters. In his excellent biography, *Dark Stream*, Leon Rousseau described Marais as growing up 'with the acute consciousness of sin and evil which typified his era, his church, his nation and his family, elements which can hurl a man from side to side like dice in a glass'.

By the age of 12 he was a published poet and was reading Shelley, Byron, Coleridge, Keats and other nineteenth-century poets. He kept a live python and stored his pens in a human skull on his desk. By 16 he had matriculated and by 19 he was the editor of a cheeky Dutch-Afrikaans newspaper, *Land en Volk*.

In 1896 Marais sailed for London to study law, but seemed to spend more time with a young woman named Susanne and rearing a baby chimp in his rooms in Lewisham. The former was to break his heart, the latter led to his first steps on a path which was to end – in the declining years of his life – with the groundbreaking study entitled *The Soul of the Ape*.

He was among a class of intelligent and articulate Afrikaners of the time who, according to the prominent poet and critic N P van Wyk Louw, had 'started dreaming in a different key'.

Dark Stream makes gripping reading, dealing as it does with the involvement of this enigmatic man with the Jameson Raid, the Anglo-Boer War, his journalism, his hypnotic effect over women, horses and chickens, his warmth towards children, his work with termites, baboons and evolution, and his struggle with morphine. But to appreciate Marais the scientist, we need a different point of entry.

In his extraordinary book *The Life of the Cosmos*, American physicist Lee Smolin has an intriguing chapter entitled simply 'What is Life?' (if you need to know the answer, read the book). In it he makes a broad but well-supported statement: 'The understanding that life on this planet is an interconnected system must be considered one of the great discoveries of science, perhaps as profound as the discovery of natural selection.'

Keep that in mind, but now let's return to Marais. In an attempt, possibly, to decrease his dependency on morphine, Marais hitched a ride on a horse cart heading for the Waterberg north-west of Pretoria, little realising its wilderness would hold him there for a number of years. It was a place, he was to write, which he'd always associated with the wonders of unpeopled veld, 'uninhabited by white men, and the wild animals were still there'.

On a farm deep in the *berge*, far from the scientific storm raised by Darwin's conclusions on the origin of species or Freud's work on the psyche, he would study two creatures which, on the face of it, had absolutely nothing in common: termites and baboons. Both fascinated him – as did all wild things.

But, insulated from the storm, Marais took a leap beyond Darwin. What was it, he puzzled, that makes termites different from baboons? And in what ways were they similar? His conclusion to both questions, as he sat in a hut amid primates and wild mountains, went beyond even Freud: the answer was memory.

Marais begun studying termites on a Rietfontein farm. On the first page of *The Soul of the White Ant* he describes this study, rather tellingly, as 'an investigation into animal psychology'.

He watched them flying from termitaria in their thousands after rain, the female flicking off her wings as she landed: 'quicker than a woman who discards her evening gown.' Then she'd raise her hindquarters and signal for a male. Having lured one, he discovered, the pair would rush round searching for a place to begin digging the first shaft of a new termitarium. The queen would grow immense, spawn hundreds of thousands of workers and soldiers, and control the hive for many years.

What Marais noticed, more importantly, was that if the pair were somehow prevented from flying, their future would end there and then. A few flaps and they could breed. Ground them, he found, and the chain of their life would be broken. They had to pass through every stage or they were doomed. His extensive study was to have two important conclusions, both pioneering but virtually unnoticed by the wider scientific community of the time.

Firstly, the whole termitarium needed to be considered as a single organism whose organs had not yet been fused together as in a human being. The queen was the brain and womb, the workers were the mouthparts and tissue builders, the soldiers acted as white blood corpuscles, and the humus gardens were the stomach. And secondly, the actions within the termitarium –

and by extension the 'hive mind' – were completely instinctive. Any variation would endanger the creature's particular, hard-won ecological niche.

Marais began writing *Soul of the Ape* in 1916 but never finished it. It was pieced together years later and published posthumously. His studies, however, began years earlier in a Waterberg devoid of guns, which had been confiscated from farmers after the Anglo-Boer War. Conditions for getting close to a troop of wild chacma baboons – which had never heard a shot fired – were ideal.

'We approached this investigation without any preconceived ideas,' he was to write. 'Although at the beginning inexperience may have left much to be desired in our methods, we had at least no theories to verify.'

But science requires theories and Marais devised a few of his own. Unlike termites, he suggested, chacmas – and by extension all primates – had the ability to memorise the relationship between cause and effect. They could accumulate personal memories and, importantly, could therefore vary their behaviour voluntarily.

This was not an evolutionary refinement of instinct but another type of mind altogether, which Marais named 'causal memory'. While the hive mind of termites was all instinct, the causal memory of the chacmas operated quite differently, virtually submerging instinctive mind. But why? Here comes the good bit ...

Natural selection, he said, was not so much the survival of the fittest as Darwin had insisted, but the line of least resistance. Those species best able to adapt to their specific environment survived more surely than those that didn't. Natural selection, therefore, had a tendency to localise, as well as specialise, species.

The downside, of course, was that sudden environmental change meant the destruction of specialists because instinctive memories only adapt over aeons of time. The solution? Listen to

Marais at his lyrical and scientific best:

'If we picture the great continent of Africa with its extreme diversity of natural conditions – its high, cold, treeless plateaus; its impenetrable tropical forests; its great river systems; its inland seas; its deserts; its rains and droughts; its sudden climatic change capable of altering the natural aspects of great tracts of country in a few years – all forming an apparently systemless chaos, and then picture its teeming masses of competing organic life, comprising more species, more numbers and of greater size than can be found on any other continent on Earth, is it not at once evident how great would be the advantage if a species could be liberated from the limiting force of hereditary memories?

'Would it not be conducive to preservation if, under such circumstances, a species could either suddenly change its habitat or meet any new natural conditions thrust upon it by means of immediate adaptation?'

What then would be the nature of the change necessary to bring this about?

'No generalising perfection of hand or foot, nor the attainment of the upright position, nor the transformation of any function in any single organ could render a species immune from the danger inherent in suddenly changing conditions.

'It is not conceivable that it could have been attained in any other way than through a modification of the brain and its functions. In other words, the attributes selected had necessarily to be psychic.'

These adaptations had allowed primates – particularly humans – to penetrate the most varied of natural environments. We became, in the words of Marais, 'citizens of a larger world'.

It is hard to believe that Marais was formulating these ideas nearly a hundred years ago. They raise important questions for psychology and the notion of the unconscious mind and they anticipate the discoveries by Raymond Dart, Louis Leakey and others, which support the thesis that the human adventure began in Africa. They also demonstrate the wisdom – later followed by work on primates by researchers such as Jane Goodall and Dianne Fossey – that the only sensible way to study wild creatures is in the wilds.

But, perhaps even more importantly, Marais's ideas were grounded on his intuition that life on this planet is an interconnected system. By the end of the twentieth century this would become a basic natural science principle.

Marais wrote *The Soul of the White Ant* in Afrikaans, in newspaper instalments, and it was only brought to wider attention by being seemingly plagiarised by a Belgian Nobel Prize winner named Maurice Maeterlinck. *The Soul of the Ape* was incomplete and, originally, published only in South Africa. Marais is remembered more for his contribution to Afrikaans literature than to science. When social anthropologist Robert Ardrey dedicated his path-breaking book, *African Genesis*, to Eugène Marais, the reaction from the world's scientific community was 'Eugene who?'.

Ardrey's reply came in an introduction to a Penguin edition of Marais's work on ants and baboons published in 1973. 'As a scientist he was unique, supreme in his time, yet a worker in a science then unborn. He was a freak, spawned by the exuberance of mankind, an immortal who speaks from his grave: beware, and do likewise. That he died by his own hand may seem as an afterthought.'

But Marais left us with a warning. Natural selection requires both a challenge and a response. To the extent that modern societies conquer the challenge, he wrote, they dilute their ability to respond.

We are no longer merely altering our particular habitat, we're altering our entire environment as well. This can bring about great changes – and dangers. Will we be able to respond as a species? Was it morphine or prescience that made Marais think it unlikely?

Caves of antiquity

'Handaxe. Banded ironstone,' pronounced Judy Maguire, flipping the rich red-brown artefact out of a Tupperware lunch box. 'Early Stone Age – about 400,000 years old. It was probably a reject; it hasn't been used.'

I picked up the glittering, 15-centimetre implement. It was perfectly teardrop-shaped and had been beautifully chipped to a keen edge all round, still sharp after all those years. It lay in my palm, the right weight, the right shape, utterly comfortable, undoubtedly made for a hand like mine. It seemed brand new: straight out of the corner troglodyte shop.

We were standing below the dump at the old Makapansgat Limeworks just east of Potgietersrus. Judy zipped open a blue holdall I'd mistakenly presumed was her overnight bag, hauled out a Checkers packet and dipped her hand into it.

'Half a fossilised brain. About three and a half million years old. It's that of a child, found at Taung. Of course it's really the sediment which half filled the braincase and solidified. Professor Raymond Dart identified it in 1925. Then he found this ... '

Judy dug into another packet and produced the front of a tiny skull. It fitted perfectly over the brain. Another packet offered up the lower jaw. The even milk teeth protruded like those of a youngster in need of orthodontics, the nose area receded and there were no ape-like ridges above the eyes.

'It was unprecedented – a human-like, ape-like creature but with a brain unexpectedly large for an ape,' commented Judy, holding the skull like Hamlet and staring into its empty eye sockets. 'Dart named it *Australopithecus africanus*, southern ape of Africa. This proto-man, he said, was bipedal, cave dwelling and predatory.

'This claim proved to be hugely controversial. Dart was denounced – both in public and academic circles – for daring to suggest humans had descended from apes. Detractors said he'd merely found an aberrant chimp and was over-interpreting the available evidence.

'Then Robert Broom found an adult skull at Sterkfontein – dubbed Mrs Ples – and Dart's claims were vindicated. And in 1947 James Kitching found *africanus* fossils in these dumps at the limestone mine.'

I'd phoned Judy a few weeks earlier for information about Makapansgat. 'Don't ask; come and see,' she'd replied. 'I'll go with you.'

It wasn't a chance I could pass up. Judy's part of an extraordinary geneaology of local palaeontologists who are among the finest in the world. They include such names as Broom, Dart, Tobias, Kitching, Brain, Clarke and Berger – people who have utterly changed our understanding of human evolution.

The Makapan Valley is part of a highlands system which tilted into existence hundreds of millions of years ago. The horizontal sediments of the Transvaal Supergroup were penetrated by a massive upwelling of igneous rock and then sagged in the middle, popping up hills on its perimeter. The northern rim of this immense soup bowl includes the Makapansgat Highlands – its southernmost rim is the Magaliesberg chain.

In between lies the seemingly endless Springbok Flats, parts of which would once have been marshy wetland supporting boundless animal and bird life. The highland rim also provided stepping-stones for forest plants, animals and humans to enable migration across the surrounding plains.

Within living memory vast, lemming-like herds of springbok traversed these Flats. Cronwright, the husband of writer Olive Schreiner, described a migration in 1925: 'Over 100 miles long by 15 miles wide was covered by the *trekbokken* moving in an unbroken mass, giving the veld a whitish tint, as if covered by a light fall of snow.'

Time and the elements have eroded the northern edge of the sedimentary uplift. The crest of this is Black Reef quartzite; now a mist belt sheltering isolated patches of relict forest and penetrated by deep limestone caves in the underlying dolomite. The wealth of usable natural resources in the area, together with the availability of protection, was to become a three-million-year-old cooking pot for primate gene experimentation – a 'hominid heartland' with its history written in the rum-tumble breccia infill covering the many cave floors.

Standing in the scrubby bush in the near-deserted valley floor, I found it hard to imagine this nursery of life. Three lime kilns, stoked on local hardwoods and burning day in and day out for more than 30 years, had put paid to the surrounding woodland. The *trekbokken* were all gone to braaivleis or biltong, the marshland had vanished, the animals had been shot out, and the rivers had dried up because of climatic changes and bad farming. The silence of progress was deafening.

But the stones still sang with their memories. Behind us, row upon row of rocks had been piled into neat walls by research students and almost every piece contained fossil remains.

Judy pointed to a mountain of debris which had been removed to get to the limestone. 'It's all fossil-rich breccia.' The place was a veritable bone-yard.

We climbed over the fossil dump and threaded our way into the innards of the hill, along a winding corridor of rock. It led into an immense cavern, held up in places by towering wood-and-cement pit props and seemingly very precarious. The miners – working from around the time of the Anglo-Boer War to the mid-1930s – had blasted out a mass of limestone, leaving a roof with heights varying from around a metre to more than 30.

Above us, embedded in what must have once been the cave floor, were fossil skulls, leg bones, teeth, blow-fly pupae and even fossilised dung balls of an ancient species of dung beetle.

The fossils of more sinister creatures have also been found there – enormous porcupines, dassies the size of large dogs, and sabre-toothed cats with jawbones which could pivot downwards to an extraordinary degree to allow food to get past their huge incisors. A large number of our ancestors must have gone down their greedy gullets.

The caves were created when rainwater combined with carbon dioxide to form carbonic acid. This became progressively more acidic as it passed through the surface soil and began corroding the dolomite sandwiched between harder bands of chert.

Cavities, and then vast chambers, were dissolved out of the dolomitic rock and filled with water. But when valleys deepened the water table dropped, draining the caverns and leaving perfect habitations for all manner of creatures, including bands of proto-humans.

Some of this dolomite contains the remains of life forms which are mind-bendingly ancient. Just outside the mouth of Historic Cave, some Malmani dolomite contains stromatolites, the fossilised remains of blue-green algae which grew in a shallow sea hereabouts nearly 2,500 million years ago.

From the limeworks we threaded our way through the bush to the Cave of Hearths, so named for the discovery of ancient fire sites. The cave offers the earliest evidence of the controlled use of fire in Africa, and viewing a Stone Age hearth with the discarded remains of a meal within its blackened circle generates an eerie feeling.

The cave fill has been extensively excavated by members of the University of the Witwatersrand staff (notably archaeologist Revil Mason and his researchers) and the labels painted on the rock face go from Early Stone Age upwards to the Voortrekkers.

From there it's a short walk to the adjoining Historic Cave, or the cave of Mokopane. It was in this huge cavern, in 1854, that more than a thousand Ndebele under Chief Mokopane starved to death while Boers maintained a siege at its two entrances.

What exactly started the hostilities is still subject to conflicting histories. Possible causes were the indenture by the Boers of orphaned Ndebele children as unpaid labourers (how they became orphaned is open to speculation), the killing of Chief Mokopane's brother by Boer leader Hermanus Potgieter, or a dispute over cattle.

There were also undoubtedly larger issues at stake, with the Boers appropriating the best land by force of arms, extracting labour through an *inboekseling* system, and appropriating grazing, ivory, game and skins.

Whatever the pretext, Potgieter and his hunting party were attacked and all but Potgieter were killed outright. The leader was dragged to the top of a hill and flayed alive. According to oral tradition his skin was used in rites for many years afterwards and is said to still exist.

The Ndebele then went on a killing spree, murdering men, women and children and dashing their brains out on two sturdy camelthorn trees at a place now named Moorddrift. The Boers mounted a punitive expedition and the Ndebele fled into the caves with their cattle and provisions, barricading the entrances and shooting anyone who was foolish enough to stick their head over the cave lip. Hermanus Potgieter's nephew, Piet, died this way and his body was recovered by a brave young Paul Kruger, later to become President of the Transvaal Republic.

Attempts were made to flush out the tribe by bombarding the cave mouth with a field cannon and rolling smoking logs into the cave entrances. The Ndebele attempted to break out by stampeding their cattle, and in this way Mokopane himself escaped, lashed to the belly of an ox. There was no water in the cavern, and when the siege ended 25 days later more than a thousand Ndebele were dead.

The view from the mouth of the cave up Mokopane's Valley is spectacular. Euphorbias and acacias dominate the foreground, and at the head of the valley a towering quartzite bluff seems to hover over an ancient forest. Was it like this when tiny *Australopithecus africanus* skulked there with crude bone clubs more than three million years ago? Did Stone Age men prowl in the forest, or Iron Age hunters sit by the sparkling river waiting for prey?

Today the area is owned by the Potgietersrus Municipality – bought, not for its historical significance, but to ensure the town's water supply. The caves themselves, which have been declared a heritage site, are curated by the University of the Witwatersrand.

But neither authority has the funds to even fence the area adequately, let alone protect the sites from ongoing vandalism. Squatters have moved into the lower end of the valley and it seems it's only a matter of time before one of the most important treasure houses of humankind – and an area of great natural beauty and heritage – is lost to browsing goats and subsistence plunder. Well, it's a valley with many histories ...

But if ever there was a place in South Africa that needs to be taken over by National Parks or declared a World Heritage Site, this is it. Judy is part of a team which has been asked to plan the management of the sites and the future of the valley and its surrounds. But there's no guarantee that their recommendations will be implemented.

Standing on the valley floor again, I could feel the tangible presence of the strange, compelling caves full of stories and old bones in the hillside above me – and 30 metres of bone-filled breccia told a rather disquieting tale. Humans, in their various evolving forms, killed to eat. But the Historic Cave suggested a less benign reason for bones.

Picking up Judy's beautiful handaxe and noting its deadly point, I wondered whether its Stone Age maker had only food-gathering in mind. Perhaps Robert Ardrey, who wrote *African Genesis*, had been right when he said our propensity to kill is what made us human.

On the other hand, looking at the jawbone of a sabre-toothed cat in the Bernard Price Institute's little museum at Wits University the following day, I reckoned that if the caves had been my home back then, I wouldn't have let the rock-hewn weapon out of my nervous grasp for a moment.

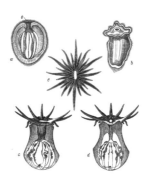

The madness of a collector

Claude Gibney Finch-Davies was not a hero, though he did bear arms and fought a war of sorts. But there was something about what happened to him which is reminiscent of the trouble the warrior Achilles had with his heel.

The Greek hero of Homer's *Iliad*, you will recall, was dipped in the river Styx by his mother in order to make him invincible. To do this she had to hold him somewhere, so she chose his heel. By not coming in contact with the river of the dead, it remained vulnerable. It was in his heel that the fatal spear entered – thrust by Paris, whose abduction of Helen had started the Trojan War. Achilles crossed the Styx for good and Paris went on to find fame and fortune.

Finch-Davies, of course, was no mythical warrior. He was born in India in 1875 to Sir William and Lady Davies. While his mother didn't literally dip him into anything, she did immerse him in natural history and her considerable knowledge of Indian snakes. At the age of six, young Claude was packed off to school in England, but wasn't too good at it. At 18 he chucked in his studies and enrolled with the Cape Mounted Riflemen in the Cape Colony. By the time the Anglo-Boer War arrived, he had made it to corporal.

The warrior business didn't really interest Davies, though he did end up with the Queen's Jubilee Medal, the King's and the Queen's South Africa Medal, a Long Service Medal, a Good Conduct Medal, the 1914–15 Star, the British War Medal and a Victory Medal – all run-of-the-mill stuff as wars go. More importantly, some time during the Anglo-Boer War, he drew a bird.

Throughout that war Davies was stationed in the Eastern Cape and is known to have explored and hunted in Pondoland and East Griqualand. His notes mention Flagstaff, St Marks, Lusikisiki, Port St Johns, Matatiele, Imboitzi Lagoon and the Tugela Mountains. Travel was by wagon or horse, so he had time to look around. His notes mention a Namaqua dove which he shot in the school grounds in Flagstaff, redcapped larks 'scarcely taking the trouble to get out of the way of one's horses', and a brown-hooded kingfisher which often sat on the crosstree of the barrack flagstaff.

The first painting in his sketchbooks, of an Ethiopian snipe shot at Lusikisiki, is dated 4 August 1903. It is amateurish but displays a good grasp of draughtsmanship and use of paint, indicating some earlier basic training in drawing and watercolour painting. These sketchbooks – there were to be 30 in all – consisted of 40 sheets, half bound and covered in canvas with a carrying strap at the end of each cover. In 1902 they cost a princely one shilling.

Soldiering soon became merely a means to an end. Davies shot, painted and ate his way through rameron pigeons ('delicious'), bitterns ('not bad eating'), yellow-billed ducks ('good sports and excellent eating') and grey-winged francolins ('undoubtedly the best game species'). Each was painted with rapidly improving skill and sensitivity, with additional meticulous notes and observations penned on the back of every sketch.

By 1906 he had completed some 200 paintings in ten volumes and, by then, his technique was so good he could make paintings of species with only the skin in his possession. A year later some of his work was published in the *Journal of the South African Ornithologists' Union*.

Davies also sent skins to the British Museum, where a previously undescribed olive sunbird subspecies was named after him: *Cinnyris olivacceus daviesi*. Fittingly, his painting of the bird was used to depict it in the museum's Bird Room.

In 1908 Davies was admitted as a member of the British Ornithologists' Union and was becoming recognised as an illustrator of international repute. He completed 69 illustrations for the authoritative *Game Birds and Waterfowl of South Africa* by Major Boyd Horsbrugh and was asked by the Duchess of Bedford to paint the ducks in her collection of waterfowl. He was by then undoubtedly the finest avian illustrator in Southern Africa, if not the entire continent.

Around 1911 Davies began devoting his time to birds of prey, eventually filling seven books with finely observed portraits. To do this he drew on specimens seen or borrowed from the South African Museum in Cape Town, the King William's Town Museum and the Transvaal Museum in Pretoria. His links with the Transvaal Museum were the beginning of what was to become both a fruitful and, ultimately, a perilous relationship.

Shortly after the outbreak of the First World War, Davies was involved in the campaign to occupy German South West Africa. He was popular with his fellow soldiers, who often brought him birds they had shot. The pockets of his no doubt less-than-crisp tunic were always full of brushes and paints, as well as occasional birds or simply pieces of birds.

He painted very fast, paying great attention to proportions and details, often using brushes with only a few fine hairs. He'd frequently forget to eat or prepare to move camp. Fortunately he had a regular batman to collect his rations, prepare camp and care for his horses.

About this time Davies began an extensive correspondence with Austin Roberts, who was in charge of the Department of Higher Vertebrates at the Transvaal Museum in Pretoria. Roberts also bought skins from him.

The young soldier's reputation was growing and it was time to look at what a man needed. South West Africa was a rather solitary place. On a trip to Cape Town, he met Aileen Singleton Finch, daughter of Captain W Finch, who was head of the Society for the Prevention of Cruelty to Animals. In August 1916 they were married. At his wife's insistence, the couple retained Aileen's maiden name and they became Finch-Davies, the name possibly appealing to Davies's avian interests.

Back in rural South West, Finch-Davies relied increasingly on books, journals and specimens borrowed from the Transvaal Museum, stressing in almost every letter how careful he would be to protect the material.

In 1918 a spat – its cause not defined in surviving correspondence – occurred between Finch-Davies and Roberts, the former complaining about being 'somewhat hurt by Mr Roberts's treatment' and suggesting that he might switch allegiance to the British Museum.

However, on being transferred to Pretoria at the end of the war, Finch-Davies donated his entire South West African bird skin collection to Roberts's museum collection on condition that he be allowed access anytime he wished.

What happened next is captured in a statement by Austin Roberts. 'Towards the end of November 1919, I happened to notice that three coloured plates had been torn from Vol 22 Cat *Birds* Brit Museum (Phasianidae) and, as I had not previously noticed this, drew Mr Sweistra's attention to it. On the 5th December 1919, I happened to refer to another work where I observed another plate was missing.'

His curiosity aroused, Roberts searched other volumes and found more than a hundred plates missing.

'Suspicion naturally centred on Lieut Finch-Davies, the only one who had the facilities of abstracting plates therefrom during the absence of the staff; yet I could not believe it possible that he would do so, from my knowledge of his care in handling books that were lent to him, and the great work he was doing in ornithology.'

The police were informed of the losses, set a trap and Finch-Davies was arrested. A closer check found 230 plates to be missing from 90 journals and books.

On 30 January the museum director, Dr Breijer, received a letter from Finch-Davies admitting the theft. 'I cannot have any excuse for what I have done ... I can only think that I must have suffered the madness of the collector, which distorts the moral sense.'

Given the seemingly out-of-character nature of the theft, and no doubt Finch-Davies's reputation, the museum authorities decided not to prosecute. Finch-Davies promised to return the plates, or substitutes for them. As security he deposited his entire collection of bird paintings – 29 volumes – with the museum until he could make good its losses.

Some strings were also pulled to avoid a court martial, and the artist was merely given a severe reprimand and transferred to the Castle in Cape Town. It could have been worse, but his hopes for promotion (after 27 years' service) were over. His wife remained in Pretoria until the birth of their third child, then moved to the married quarters in Cape Town.

On 18 May 1920 Breijer received a letter from L Peringuey,

director of the South African Museum in Cape Town, on 'a matter of great importance.' More than 130 plates had been found to be missing from the museum's bird journals and books.

Peringuey had met Finch-Davies, and naturally knew of the Pretoria incident. It appears the director made an appointment to see him. Finch-Davies never made the meeting. Early that morning an orderly took coffee to his room in the Castle and found him dead on his bed. The newspapers reported the cause of death as 'angina pectoris', but soon afterwards rumours began circulating that the real cause was an overdose of cocaine.

The rumour could not be substantiated and Finch-Davies was given a military funeral and buried at Maitland Cemetery. His wife and children, virtually penniless, returned to Ireland.

There is a strange sequel to this sad tale. In 1940 Austin Roberts published *The Birds of Southern Africa*, now affectionately known as *Roberts' Birds*. It soon became the top-selling book of its kind. Its first edition sold out in six weeks and it is about to go into its seventh edition.

Such an undertaking required clear pictures, a job given to a draughtsman named Norman Lighton, seconded to the Transvaal Museum from the Public Works Department.

In the vaults of the museum lay some 600 exquisite paintings by Finch-Davies, their ownership undecided. Austin Roberts, aware of the predicament of the dead artist's wife, had tried to get the museum to buy them, but nothing came of it.

Lighton, under pressure to produce 56 plates containing watercolours of 1,032 birds, hauled out the Finch-Davies paintings and copied them – in most cases poorly. In a biography of Austin Roberts, Bob Brain says merely that Lighton 'made extensive use' of Finch-Davies's work. Dr Alan Kemp, who has done much to bring the work of Finch-Davies to public notice, is charitable about the matter, saying that, given the body of work available to Lighton, the use of Finch-Davies's paintings was understandable.

But a close study of the work of both artists tells another story. Bird after bird is exactly copied, but certain elements – a feather here, a foot position there – are changed. Sometimes the bird's position is reversed, but then the copy is often more exact. Was this done out of personal taste or was it an attempt at concealment?

The use of all those pictures of uncertain ownership may be excusable had the original artist received due recognition. But the introduction of the first and later editions merely states: 'the plates have been figured in colour by Mr Norman C K Lighton under the directions of the author.' Finch-Davies is not mentioned.

His wife, assisted by Roberts, managed to sell two paintings still in her possession for £17 7s – the only sum she ever received from her husband's enormous artistic labours. She remained dirt-poor for the rest of her life.

On the other side of the River Styx, Finch-Davies must be regretting his light-fingered foibles in an otherwise unblemished career. Without it he would probably have been the illustrator of *Roberts' Birds* – and undoubtedly have become the most celebrated avian artist in Southern Africa. Fate works in curious ways.

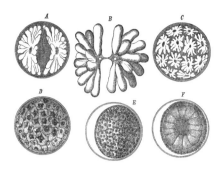

I look your mouth and it be sweet

Why are there virtually no great women travel writers? In any good bookshop you can peruse the lonely meanderings of Paul Theroux, the mystical ecology of Barry Lopez, the outrageous adventures of Tim Cahill, Bill Bryson and P J O'Rourke, the whimsical journeys of Graham Greene or the more scientific wanderings of David Quammen. Apart from the funny but often depressing Irish writer Dervla Murphy and Isak Dinesen's 'faarm in Aafrica', no women stand out.

It doesn't matter how far back in time you go, men seem to have done the travelling and written the books. Their wives appear to have stayed at home wringing their hands and wondering if their hero would return.

The ships' logs of men such as Vasco da Gama, Bartolomeu Dias, Walter Raleigh and James Cook were soon transformed into bestsellers, as were the diaries of the explorers such as

Marco Polo, Ibn Battuta, Richard Burton, François le Vaillant, Robert Scott, David Livingstone and Henry Morton Stanley. Women remained silent. Even the imaginary, fantastical *Odyssey* and *Gulliver's Travels* by Homer and Swift – which you could stay at home and write between nappies – were penned by men.

Home-bound, weather-sodden Europeans couldn't get enough of these sun-drenched Otherworlds of exotic beasts, bizarre landscapes and pointy-toothed cannibals. Would they have shunned a woman adventurer?

One woman stands proof that at the end of the nineteenth century, at least, they didn't. And what a champion of her sex she was: that irrepressible, wickedly witty, outrageously daring, undeniably eccentric Victorian spinster, Mary Kingsley. She did for fun travel writing what the Box Brownie did for photography – and had stuffy Victorians popping waistcoat buttons and splitting whalebone corsets with laughter. But who knows of her today? Her finest book, *Travels in West Africa*, is mostly buried in dusty research libraries under 'African exploration'.

When her first book, *Travels*, was published in the 1890s it was an instant success – fuelled, one suspects, by the incongruity between Mary's appearance and her adventures. On her lecture tours to publicise the book, people were nonplussed by the prim lady in a dark, full-skirted dress, buttons to her chin, an old-fashioned pillbox hat and her hair in a tight schoolmarmish bun discoursing about smacking crocodiles with a paddle, cresting mountains where men had failed, and the virtues of cannibalism.

Her exploits went right off the register of what a nice young lady of her generation did. She lived in an age when women's subjugation was so complete that while her parents sent her younger brother to Cambridge University, they didn't consider giving her a formal education at all.

Her father, George, had got his cook pregnant and, to prevent a scandal, married her. Two weeks later Mary was born. But married life didn't settle with his ambitions. He was a doctor, and spent most of his days travelling the world attending to highborn Englishmen in the South Seas (and, it was rumoured, fornicating with Maori maidens). He left his family comfortable but mostly without him.

Kingsley stayed at home and ministered to her increasingly sickly mother. She had few friends, seldom went out, but had the run of her father's extensive library. There she cosseted herself with works such as Mungo Park's *Journal of a Mission to the Interior of Africa*, Richard Burton's *Abeokuta and Cameroons Mountains*, a favourite by Johnson, *Robberies and Murders of the Most Notorious Pirates*, and a monthly journal *The English Mechanic*. By the time she was 13, Stanley was negotiating the rapids of the Congo River.

When her father returned she revelled in his romantic stories, admiring his adventurous spirit but resenting him for curtailing her own. Despite her academic disadvantage she befriended Cambridge professors and lecturers. One was to write, later, that Mary 'belonged to the order of native-born genius which cannot be classified'. Her wandering father and her uncle, Charles Kingsley, author of *The Water Babies*, were probably inspirations. Or maybe her fighting cock named Attila.

In 1893, when she was in her late 20s, her father and then her mother died within a few months of each other.

'For the first time in my life,' she was to write, 'I found myself in possession of five or six months which were not heavily forestalled, and feeling like a boy with a new half-crown, I lay about in my mind as to what to do with them. "Go and learn your tropics," said Science. Where on Earth am I to go? I wondered. So I got down an atlas … '

With a vague intention to finish a book her father had been writing, Kingsley booked a passage to West Africa on the cargo tramp SS *Lagos*, packed her portmanteau, parasol, cutlass and notebooks and, with a rustle of petticoats, was gone. When asked on board why she was going to the tropics – widely considered to be a 'white man's grave' – she answered that it was to pursue an interest in 'fetish and fish'.

On 17 August 1892 she was lowered from the steamer into a surfboat off Calabar. Mary Kingsley, the African traveller, had just been born. 'Out of my life of books', she wrote delightedly, 'I had found something.'

Back on the *Lagos* a few weeks later and up the Bonny River (in present-day Nigeria), she wrote to a friend in a style which was to deepen into wisdom and twinkle with wit:

'Here I am on my first river, but not my last as long as I have life and money to get there. I am already formulating methods of economy for my London use – not a theatre, nor an extra omnibus fare, nor an extra garment until I smell again the heavy rank land smell, see the blue ocean turn cocoa colour in a sharp line and hear the music of thunder across the Bonny bar.'

By December she was back in Britain, promenading with a monkey perched on her shoulder. Several months later, true to her word, she was back on the malarial coast.

Kingsley was to undertake two extended trips to West Africa, the second with the blessing of the British Museum of Natural History.

She wrote extensively on primitive fetish, had three new species of fish named after her, and brought back 'one absolutely new snake and one lizard the British Museum has been waiting for for ten years'. But her real contribution to science – and to travel writing – was her lyrical, fine-textured descriptions of places few Europeans had ever seen.

Mary travelled, with no support or government backup, and often with only a native guide, up the crocodile-lined Congo and Ogowé rivers, through jungles (spying on gorillas and sleeping in cannibal villages) and climbed the 4,000-metre Mount Cameroon (alone, eventually, because her guides couldn't stand the cold).

Time drifted. She confessed to a friend in a letter dated 'July or August' but written in the middle of May: 'It is one of my disastrous habits well known to my friends on the Coast that whenever I am happy, comfortable and content, I lose all knowledge of the date, the time of day, and my hairpins.'

She wore – in canoes, squelching through mangrove mud or stumping up a mountain – dark, full-length, many-petticoated dresses, high-necked bodices, pillbox hats and carried a parasol. There were trips, however, when she tucked a loaded pistol and a cutlass into her belt.

It is not my intention to summarise Kingsley's travels. Track down her two books and don't fail to read them. I wish, only, to tantalise you with the delicious evocativeness of her writing about raw nature, and to suggest that a woman's-eye view is somehow different from a man's.

'I should like you to see the altars they have in their yards with human skulls sprinkled with fresh blood daily,' she wrote of coastal villagers. 'Nothing is known by white men of the people behind these coastal tribes, but the native traders tell of great inland towns surrounded by cultivated rich land and reeking in human sacrifice.'

Thus tantalised, she boarded a river steamer and headed inland into the epitome of Joseph Conrad's *Heart of Darkness*. 'All West African steamers', she told her readers, 'have a mania for bush, and the delusion that they are required to climb trees. The *Fallabar* had the complaint severely, because of her defective steering powers and the temptation of the magnificent forest, and the rapid currents, and the sharp turns of the creeks offered her. She failed, of course – they all fail – but it is not for want of practice. I have seen many West Coast vessels up trees, but never more than fifteen feet or so.'

Upriver, Kingsley employed two virtually naked local men to pilot her canoe. She packed a net, fishing lines, paddles, some bait in an old boiled-mutton tin, and a calabash to use as a baler. Then, tucking a *Study of Fishes* by Günther under her arm, she set off up the Ogowé River to explore.

Within two hours of leaving they were facing their first rapid. It did not deter Mary; she merely marvelled: 'Great grey masses of smoothed rock rise up out of the whirling water in all directions. The effect produced by this, with the forested hillsides and little beaches of glistening white sand, was one of the most perfect things I have ever seen.'

Later, overnighting in a village of Fang cannibals, she slipped out in the small hours to contemplate the great river.

'In the darkness round me flitted thousands of fireflies and out beyond this pool of utter night flew by unceasingly the white foam of the rapids: sound there was none, save their thunder. The majesty and beauty of the scene fascinated me, and I stood leaning with my back against a rock pinnacle watching it.

'Do not imagine it gave rise, in what I am pleased to call my mind, to those complicated, poetical reflections natural beauty seems to bring out in other people's minds. It never works that way with me; I just lose all sense on human individuality, all memory of human life, with its grief and worry and doubt, and become part of the atmosphere. If I have a heaven, that will be mine.'

Kingsley learned to paddle a dugout canoe and went exploring the great mangrove swamps. Being tidal, the water drained out, once leaving her stranded in a muddy pool full of crocodiles. With typical Kingsley sang-froid she described what followed:

'A mighty Silurian chose to get his front paws over the stern of my canoe and endeavoured to improve our acquaintance. I had to retire to the bows, to keep the balance right, and fetch him a clip on the snout with a paddle, when he withdrew. I should think that crocodile was eight feet long ... a pushing young creature who had not learned manners.'

Kingsley later enlisted four guides to take her through unexplored jungle controlled by the Fang. The guides agreed to go because of the pay and because, they told her, 'they look my mouth and it be sweet, so palaver done set'.

They negotiated accommodation in a Fang village and Kingsley, exhausted from the day's trekking, curled up among the boxes in a hut with her head on a tobacco sack and dozed.

'Waking up I noticed the smell in the hut was violent, from being shut up I suppose, and it had an unmistakably organic origin. I investigated, and tracked it to those bags, so I took down the biggest one and shook its contents into my hat, for fear of losing anything of value. They were a human hand, three big toes, four eyes, two ears and other parts of the human frame. The hand was fresh, the others only so so, and shrivelled. I subsequently learned that although the Fang will eat their friendly tribesfolk, yet they like to keep a little something belonging to them as a memento.'

Cannibalism, she concluded – shocking her Victorian readers – was no danger to white people 'except as regards the bother it gives one in preventing one's black companions from being eaten'.

Near the village of Efuoa, Kingsley found good justification for her voluminous Victorian garments. 'I made a short cut and the next news was I was in a heap, on a lot of spikes, some fifteen feet or so below ground level at the bottom of a bag-shaped game pit. It is at these times you realise the blessings of a good thick skirt. Here I was with the fullness of my skirt tucked under me, sitting on nine ebony spikes some twelve inches long, in comparative comfort, howling lustily to be hauled out.'

Her advice for all travellers? 'Never lose your head.'

Rejoining the Rembwé River, Kingsley hitched a ride back to the coast. By then an accomplished canoeist, she was handed the tiller and piloted the dugout through the night as its crew slept.

'Much as I have enjoyed life in Africa,' she wrote of the experience, 'I do not think I ever enjoyed it to the full as I did on those nights dropping down the Rembwé. The great, black, winding river with a pathway in its midst of frosted silver where the moonlight struck it: on each side the ink-black mangrove walls, and above them the band of star and moonlit heavens that the walls of mangrove allowed one to see. Forward rose the form of our sail, idealised from bedsheetdom to glory; and the little red glow of our cooking fire gave a single note of warm colour to the cold light of the moon.'

Homeward bound, Mary decided to stop over in the Cameroon to climb Cameroon Mountain – the 'Throne of Thunder' climbed by Burton but few others, and never by a woman.

'Now it is none of my business to go up mountains,' she counselled herself. 'There's next to no fish on them in West Africa, and precious little good rank fetish ... ' But the temptation proved too great and off she went, parasol in hand.

Kingsley returned to Cambridge to produce her second book. By 1900 she was ready to revisit Africa when the Anglo-Boer War broke out, so she headed for Cape Town 'to be of some little use'. She installed herself in the British Hotel in Simon's Town and volunteered as a nurse at the Palace Barracks Hospital.

There she tended Boer prisoners, dragging men delirious from typhoid back to their beds by their shirt tails and heaving those who had fallen to the floor in a fever back between their sackcloth sheets.

It was barely eight years since Mary's mother had died and here she was again, tending the dying. In those eight years she had written two books and become a controversial figure in British colonial politics, campaigning against hut tax and liquor prohibition. She had also become a first-class travel writer.

'It never occurs to me', she was to write, 'that I have any right to do anything more than now and then sit and warm myself at the fire of real human beings.'

Several weeks after arriving in South Africa, Mary Kingsley contracted typhoid and died. She was 37. She had requested to be buried at sea, so her body, was encased in a lead-lined coffin and – with full military and naval honours – consigned to the deep waters of False Bay, off Cape Town.

'Why did I come to Africa? Thought I. Why?' she had once asked herself. And answered: 'Who would not come to its twin brother hell itself for all the beauty and the charm of it!'

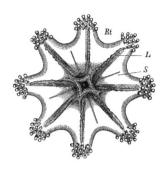

Some lessons on dreaming the world

You can still change the world by dreaming it,
You still have your tricks, old unteachable, untameable;
You could still make an eland from a piece of old boot,
You could still create the moon from an old, bent shoe.

!Xam Bushman song

The virtual elephants took small trips initially, the blue males ranging a bit further than the yellow females. Then a male – blue and square – began patrolling the park boundary and, after some hesitation, streaked across no-ellie-land to another park.

In case you think I've just smoked my socks, all this was happening on a screen under the direction of Iain Douglas-Hamilton, Kenya's elephant man. A number of elephants had been fitted with satellite trackers and their movements were represented on his computer by little moving squares.

An interesting pattern began to emerge. The creatures had

places of gathering and favoured corridors. Over an entire season the pattern of their movements had shape and meaning. Iain's aim is as magnificent as it is obvious: to let the elephants show him where they want to roam, then to attempt to redraw Kenya's park boundaries accordingly. It's decidedly unusual for people to ask animals what they want.

When I mentioned this plan to Cormac Cullinan he became quite excited. Cormac's an environmental lawyer who got round to wondering why all laws treated humans as subjects and everything else as objects.

'The only rights recognised by law are those enforceable in a court of law,' he pointed out, 'and these may only be held by human beings. This means that, from the perspective of our legal systems, the millions of other species on the planet are outlaws. And that's precisely how we treat them.'

That sort of stuff's not going to get him elected to the Law Society, but he had a point. The idea started a trail which led to the feet of an extraordinary Catholic monk named Thomas Berry.

Berry, born in America of Celtic ancestry and now well in his eighties, entered a Catholic monastery in 1934. He was, he confessed, a broody lad, and figured out there were only two places to brood: jail or a monastery. He chose the latter.

After writing his doctorate on Western intellectual history, he studied Mandarin and Sanskrit, becoming deeply attracted to Confucian and Buddhist thought. Some 20 years ago he moved from cultural to planetary history, becoming – as he puts it – neither a theologian nor a historian but a geologian.

He's a spine-tinglingly intelligent man; undoubtedly one of the foremost cosmic thinkers of our age. His ideas need something of a deep-time preamble.

The timescale of earth is divided into eras, Precambrian beginning about 800 million years ago, then moving up to the present through Palaeozoic, Mesozoic and Cenozoic.

In the 70 million years of the Cenozoic Era the story of life on this planet flowed over into what could be termed the lyric period of earth history. The trees had developed nearly 200 million years before this, the flowers had appeared perhaps 30 million years before, mammals already existed in rudimentary form.

But in the Cenozoic Era there was wave upon wave of life development, with the flowers, the birds, the trees and the mammal species, particularly, all leading to that luxuriant display of life upon earth such as we have known it. All without human inteference. Towards the end of the Cenozoic, our ancestors arrived on the scene, but for most of their three million or so years they existed alongside the rest of creation without interfering much.

Then, in the late eighteenth century, we began the wide-scale use of fossil fuels – and with this came the tools and social conditions to utterly alter the planet's biological and geographical history.

How did we do this? Well, take carbon. It's a magical element, present in all living things, but too much of it in the wrong places is a bad thing, as anyone downwind of a factory will know. The earth has immense quantities of the stuff, but nature achieved a safe balance by storing it in coal and oil deposits and in the great forests. By liberating it through burning, according to Berry, humans have profoundly affected the rhythms of the atmospheric world. Then there's river pollution, sea pollution, mono-crop production, ozone depletion, habitat destruction, mass species extinction ...

While we have achieved magnificent things, we've stacked up deadly problems at the planetary level. One of Berry's interests is what he terms 'macrophase biology', the integral functioning of entire complex biosystems. In the twentieth century, for the first time, the industrial context of human society was a defining element in the functioning of other life systems on the planet.

According to Berry, this is a time of planetary depletion of resources in which humans are the primary agents of biological history. It's a period in which extinctions are taking place on a scale unequalled since the terminal phase of the Mesozoic Era (which saw the end of the dinosaurs). Some estimates place the loss of species at around 10,000 a year.

'We are upsetting the entire Earth system,' writes Berry, 'which over billions of years and through an endless sequence of groping, of trials and errors, has produced a magnificent array of living forms, forms capable of seasonal self-renewal over vast periods of time. Most amazing is the inability of our religious or educational establishments to provide any effective religious or ethical judgement on what is happening.

'The natural world surrounding us is simply regarded as the context in which human affairs takes place. In the presence of humans, the natural world has no rights. We have a moral sense of suicide, homicide and genocide, but no moral sense of biocide or geocide, the killing of the life systems themselves and even the killing of Earth.

'We find ourselves ethically destitute just when, for the first time, we are faced with the irreversible closing down of the Earth's functioning in its major life systems.'

Both science and religion, Berry claims, suffer from a 'dysfunctional cosmology'. Unless we learn that the universe is a communion of subjects, not a collection of objects, that our actions and laws are subordinate to those of the planet, we are heading into an evolutionary cul-de-sac. The result, he warns, will be survivors – if there are any – 'living with all their resentments amid the destroyed infrastructures of the industrial world and amid the ruins of the natural world itself'.

For Berry, the solution lies not in science or technology, but in a moral, ethical, legal and spiritual re-tracking. For millennia life systems have worked on balance, where no life system utterly overwhelms any other. But our technologies have enabled us to transcend this situation. We have no limit. So the basic recipe for our survival has to be self-limiting with regard to resources, habitat, population and consumption. We have to learn – by fear if necessary – that everything lives in and by everything else. This period (if we don't choose a destructive 'solution') Berry has termed the Ecozoic Era, a time in which harmony with nature is restored by force of will.

In the period we're in, the monk insists, the religions of the world have lost the power to awe. We are – in these final years of the Cenozoic Era – 'between stories'. What we need is a new Great Story to heal our rift with the spirit of nature.

'No one ever before could tell, in such lyric language as we can now, the story of the primordial flaring forth of the universe at its beginning, the shaping of the immense number of stars gathered into galaxies, the collapse of the first generation of stars to create 90-something elements, the gravitational gathering of scattered stardust into our solar system with its nine planets, the formation of the Earth with its seas and atmospheres and the continents crashing and rifting as they move over the asthenosphere, and the awakening of life.

'Such a marvel is this 15-billion-year process; such infinite numbers of living things on Earth, such limitless variety of flowering species and forms of animal life, such tropical luxuriance. Now we are experiencing the pathos of witnessing the desecration of all this.

'We urgently need to tell this story, to meditate on it, and listen to it as it is told by every breeze that blows, by every cloud in the sky, by every mountain and river and woodland, and by the song of every cricket ... '

There are many who are putting flesh on Thomas Berry's Great Story: Iain Douglas-Hamilton giving elephants a voice, Cormac Cullinan puzzling over a new jurisprudence for all earth, far-sighted people creating huge transfrontier parks in Africa, private individuals turning their land into wilderness, selfless biologists attempting to save all manner of flagging species, activists challenging George Bush's planned destruction of pristine Alaskan snowfields in search of oil.

People looking for alternatives, however, are realising that the answers may come from the remnants of older cultures: the Bushmen, Amerindians, Aborigines and others; people who, for thousands of years, learned to live in harmony with nature. They have, says Berry, the ingredients of the new Great Story. We need to cherish them and learn their stories with humility before they are lost forever.

'Remember, Little Cousin,' a Bushman elder is recorded as having said to a young boy, 'no matter how awful or insignificant, how ugly or beautiful it might look to you, everything in the bush has its own right to be there. No one can challenge this right unless compelled by some necessity of life itself.

'Everything has its own dignity, however absurd it might seem to you, and we are all bound to recognise and respect it if we wish our own to be recognised and respected. Life in the bush is necessity, and it understands all forms of necessity. It will always forgive what is imposed upon it out of necessity, but it will never understand and accept anything less than necessity.

'And remember that, everywhere, it has its own watchers to see whether the law of necessity is being observed. You may think that deep in the darkness and the density of the bush you are alone and unobserved. But that, Little Cousin, would be an illusion of the most dangerous kind. One is never alone in the bush, one is never unobserved.'

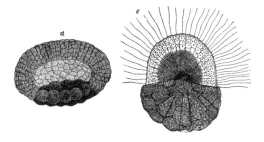

The river makers

His job, he said, was making rivers flow. In any other context Michael Alla may have been certified as having a God complex or being stoned out of his gourds. But sitting on a hillside beside the Palmiet River in the wild Outeniqua Mountains, he had proof. The river certainly was flowing, and all around this quiet man were the signs of the five years of work that had made it happen. Further downstream, though, the effect had not yet filtered through. It was peak holiday season in Plettenberg Bay and a large sign strung across the main street read 'WATER CRISIS. There is only domestic water available. No watering of gardens permitted.' The really big problem, it seemed, was the beach showers. They'd been cut off and visitors were pretty upset.

'I mean, how can we get the sticky salt water off us after we swim?' one woman complained. 'This water thing is really unacceptable.'

She had a point, even as she missed it: one of the most extraordinary things about water is that its molecules are sticky, which is why water forms drops rather than dispersing all over the place.

Its molecules not only have a tendency to cling to each other, they grasp at external surfaces, drawing the entire body of water along behind them. They can even move uphill, rising up capillaries to incredible heights. Among other things, this special talent makes trees possible – which is fine if you like them, but bad if they're invasive aliens.

In fact water breaks all the rules. Consider: according to the laws of physics, substances become denser as they get colder, shifting from gas to liquid and finally to a solid state.

Most of them do – but water only sticks to the rules until it reaches 4°C. Then it does something strange: instead of shrinking, it starts to expand. By 0°C – when water freezes – it has grown in volume by ten per cent. That's why rocks split, forgotten bottles of wine burst in the freezer and water pipes crack open on cold Highveld nights. It's also why there's life on earth.

If water didn't do this, ice would sink. Think about that. Soon every ocean would be solid – closely followed by dams, then pools – each ice layer freezing the one above it. Earth's water would be locked away; springs would not spring, streams would not stream, and rain would vanish from the skies. All that ice would reflect the sun's rays back into space and we'd be a silent iceball. Give thanks for the quirkiness of H_2O.

But water has even more tricks up its sleeve. In *Dreams of Dragons*, biologist Lyall Watson points out that because everything is essentially 'chemical', and because nearly every chemical reaction requires an aqueous solution, water can combine with and therefore corrode almost everything. It can wear down mountains and chew through the hardest steel.

Without this quality, however, no essential mineral could pass from soil to roots or flow up a plant's stem to its flowers. No animal could digest food or transport nourishment in its bloodstream. Leaves wouldn't be able to absorb life-giving gases and we couldn't take in oxygen through our lungs. Life would be impossible.

Of course we don't only use water, we're made of the stuff. Most living things are waterlogged. Jellyfish consist of about 95 per cent water, frogs 80 per cent, chickens 75 per cent, and we check in at around 65 per cent. We bottle up around 38 litres of the stuff and have to replace at least two of them a day. Herein lies a problem.

It is generally agreed that water is an uncommon substance in the cosmos, but on earth we're particularly well endowed. How they do this I don't know, but it's been calculated that the total supply of water on earth is 1,335 million cubic kilometres. The trouble is that 97 in every 100 litres are in undrinkable oceans and a further 2 per cent (29 million cubic kilometres) is locked up in polar caps and glaciers. That doesn't leave much: just over half a per cent lurking around in groundwater, 0.009 per cent in freshwater lakes and 0.0001 per cent in rivers.

So when you do the sums, almost every living thing not flipping round in the oceans is dependent on one per cent of the earth's water – and most of that is underground anyway. So the dead showers on Plettenberg Bay beach are a symptom of a problem much larger than sticky water. Which brings me back to Michael Alla.

I'd been nosing round the water problem when Guy Preston of the Working for Water Programme called to say I should visit one of his projects.

'They're doing good work,' he said. 'You should go take a look.'

A few months later I found myself bouncing down what passed for a road to the Tsitsikamma hamlet of Soetkraal.

Working for Water ecologist Pam Booth was wrestling with the bucking wheel and I fervently hoped she could talk and drive at the same time because it was a long way to the valley floor. In fact, she was holding forth with an enthusiasm and pride seldom seen these days in ecologists, whose job seems to be to tell us how fast we're wrecking our planet.

'This job does it for me,' she was saying. 'I've never come across a project that gets so many things right. We're creating jobs, conserving the environment and creating water security for the whole region. Do you know what it feels like when a stream starts to flow again after we've yanked out the invading alien trees? You don't realise how precious water is until it's gone. Here we're bringing rivers back to life.

'There's Michael Alla and his team. Let's go talk to them.'

So that's how I met the man who makes rivers flow.

'First we take out the underbrush, then we take out the wattles,' he told me, leaning on a particularly mean-looking slasher. 'We take out pines, the wattle, the hakea. They can steal a whole river. When we finish with an area the fynbos starts to grow and the streams flow. A few years later we have to go back because the seeds have made more trees. It's hard work, but just look at that river ... '

Michael began as a labourer and – like many of the 20,000 or so others in Working for Water projects across the country – was taught to use a bush cutter and a chainsaw. He proved to be pretty smart, learned alien identification and herbicide use. Now he's a team boss, about to become an independent contractor to the Tsitsikamma project, and has trained to be a certified field guide. In his spare time he specialises in nature walks and has discovered several Bushman rock-art sites.

Some of his team are being taught to read and write, others have picked up skills in permaculture. The Working for Water village of Soetkraal has about 140 residents and they're all involved in alien eradication – and, indirectly, bringing the beach showers in Plettenberg Bay back to life.

Over a tuna sandwich on homemade bread and a cup of tea in Soetkraal's rudimentary communal kitchen, Pam disclosed the extent of the programme with the flourish of a new recruit to conjuring.

'In South Africa about ten million hectares are invaded by alien plants – that's about the size of KwaZulu-Natal. Since it started up five years ago, Working for Water has cleared – let me get this right – 800,000 hectares. At last year's prices it cost R750 a hectare.'

She obviously took her figures very seriously; I guess when you have to justify the programme spending hundreds of millions of rands a year, you need to. But I could see why she was excited about her job: it was eminently justifiable. More than half the people employed are women, thousands of days have been spent training formerly unskilled labourers, and all this is taking place in more than 300 areas of the country.

Since the programme began, an estimated five billion invading alien trees have been chopped down or yanked out the ground. Can you comprehend that? A single mature alien tree can consume 50 litres of water a day: a dense stand can gulp around two million litres a year for every hectare. If you multiply that by the number of hectares cleared, Working for Water is saving about nine Hartbeespoort Dams a year. Just roughly, of course, but you get the picture.

By the time we bounced back out of the Soetkraal valley I was a convert. If you take into account riverine restoration, job creation, employee empowerment and water saving, the Working for Water campaign must be one of the most effective environmental projects on the planet.

What's interesting is that people like Michael Alla, down on the ground with a slasher in his hand, know that. Michael may have started out as an unskilled labourer, but he now understands what it means to bring a river back to life. And he's helped to make it happen. There aren't many people in the world who can claim that.

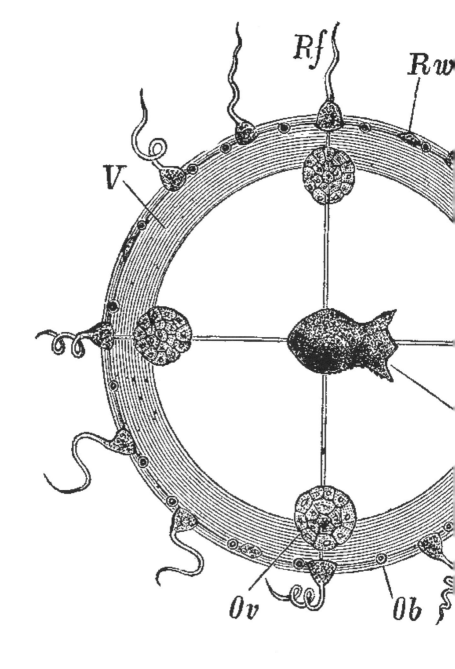

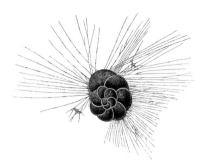

Galactic ecology

'When the end of the world happens you can come to Sutherland to get an extra two weeks,' chuckled Rudi Blom, sliding a beer to me across the upturned cable drum.

He ambled over to the fire and prodded the coals, then rolled into a complicated story about a confidence trickster, a tipsy cop and a stolen car. The shed we were in had been used to boil up soap and the fireplace consisted of a huge mud chimney. Lethal-looking gin traps for snaring jackals hung on the rough mud walls.

The door of the shed was a gaping hole into the moonless Karoo night. With black nothing overhead, the endless semi-desert in every direction and a mere handful of people in the silent village, I had a strong sense of being in the middle of nowhere. Rudi's guesthouse was the only place I'd found to stay. Its style was early twentieth century: dark-varnished furniture, an ancient valve radio and animal-skin mats.

My reason for being there was 18 kilometres out of town on the top of an utterly isolated hill. After a drink and a sausage doused in tomato sauce, I headed up the dirt road to the dark, silent domes.

Just before the big white wall bearing the name 'South African Astronomical Observatory', a beautiful bat-eared fox trotted into the road. I slammed on the brakes, wondering – in that moment which crisis gives you – whether it was worth plunging sideways off the road for a fox, and caught a glimpse of its startled eyes as my car slewed past it. Silly damn creature!

The domes, when they appeared in my parking lights (headlights were forbidden), looked huge and forbidding in the weak starlight. I could hardly find the observatory I'd been directed to. Opening the door, I was assailed by multiple sensations: the sight of a huge gantry silhouetted against the open slot in the dome and the exquisite Irish sounds of *Lilting Banshee*. Astronomer Dave Laney was in residence capturing the infrared emissions of cepheid variables.

'They're distance calibrators,' Dave explained when I looked blank. 'Cepheids are unstable stars which pulsate. I'm using them to calculate the depth of deep space.'

The explanation which followed soon lost me. Dave is one of the more understandable astronomers in South Africa, but discussing deep space requires some hefty terminology.

'It's what I call the tyranny of the very large,' he grinned when I threw up my hands in despair. 'But we've got all night.'

There are many side roads in twenty-first-century cosmology – often hotly contested – but the main highway, it seems, is to the 'big bang' and back. The road there is about time, size and numbers, the way back is pure wonder.

The numbers astronomers routinely deal with are truly mindboggling. The sun's light takes eight minutes to reach earth and only a few hours to reach the outermost planets of our solar system. Light from the nearest stars in our Milky Way galaxy has taken centuries to reach us. Which is not surprising, if you consider that if earth was the size of a sugar lump, the nearest star would be 1,600 kilometres away.

With Sutherland's telescopes astronomers can see stars and galaxies up to ten billion light years away, but because light takes that time to reach us, they're also looking back ten billion years in time. What they see are 'fossil' galaxies as they were when the universe was a mere one-tenth of its present age.

The earth is not as old as the universe, of course, but it's been around for a good few years as well. If Africa had existed forever and you were walking from Cape to Cairo, starting at the planet's birth and arriving when earth is engulfed by the sun, you'd have to take one step every two thousand years. Right now you would be about halfway: somewhere near Rwanda. A mere three or four steps would represent all of recorded history.

The number of objects out there also takes some getting used to. All but the most powerful earth-based telescopes will show the night sky as consisting of glowing objects against a black backdrop. But the Hubble Space Telescope, operating beyond the fuzziness of the atmosphere, has photographed the amazing truth: there is no part of the sky which is not filled with closely packed – though immensely distant – galaxies. It's estimated that there are more galaxies in the universe that all the stars ever seen by earth-based telescopes – and each one contains billions of stars.

Between 10 and 20 billion years ago, if the current consensus is correct, all of this spewed from a single point in empty space smaller than the size of a golf ball. Astronomers are eccentric types given to odd ideas, but these days there seems little dissent about this event – which has been dubbed the 'Big Bang'.

Dave changed the CD to something more upbeat and chuckled at my glazed expression.

'What we need is more power,' he commented as the towering telescope peered up at his cepheid variables. 'This has a 0.75-metre mirror, SALT's will be ten metres across.'

SALT stands for South African Large Telescope, which will soon go into operation beside Sutherland's present domes. It's the biggest scope in the southern hemisphere, an eye on the sky with a special watching brief on the Magellanic clouds.

The road back from the Big Bang is so finely tuned you'd swear the universe knew we were coming. Consider this: with nothing but endless, empty vacuum on all sides, the exploding 'stuff' of the Big Bang would tend to expand evenly, filling space seamlessly with an almost featureless gas of neatly spaced atoms.

Such a universe would remain featureless and dull for billions of years: no galaxies, therefore no stars, no chemical elements, no complexity and certainly no people.

But an infinitesimal moment after the bang, a 'ripple' occurred in the process. If the ripple had been too small the forces of expansion would have flattened it out; if it had been too large the result would have been unimaginably destructive turbulence. But the ripple was just enough to tear the fabric of hurtling matter, creating space between the atoms and causing clusters of hydrogen and helium which gravity coalesced into galaxies and stars. Space was born.

If you think this sounds far-fetched, let me report that the afterglow of this cosmic fireball – a microwave hiss – was detected at the Bell Telephone Laboratories back in 1964 using a sensitive antenna. Then in 1990 a satellite built to study this background radiation detected the ripple, an event heralded by the celebrated British physicist Stephen Hawking as 'the greatest discovery of the century, if not of all time'.

I mentioned that gravity gathered the gases to manufacture galaxies. Cosmologists know what gravity does, but they still don't know what it is. If it had been too weak, everything would have drifted apart; if it had been too strong, nasty collisions would be too frequent to enable stable galactic and stellar systems – and stars would have been too small to support usable planets. But, somehow, it was just right.

If all that still sounds like obvious stuff, try this: atoms are hardy things, no natural terrestrial processes can destroy them. In the beginning all was helium and hydrogen, which, fortunately, made rather good star material. But life as we know it needs more elements than that – 92 sounds about right. Even our sun couldn't cook up all that, but the cores of bright blue stars such as those in the Orion nebulae – and the intense shocks when they finally exploded – can.

When a big star has consumed all its fuel, catastrophic infall compresses its core to the density of an atomic nucleus, triggering a colossal explosion which blows off its outer layers. These pyrotechnics can outshine an entire galaxy.

The debris of such a supernova contains the outcome of all the nuclear alchemy which kept the star shining over its lifetime. There's a good deal of oxygen and carbon in this mixture, plus traces of many other elements. The calculated 'mix' is almost exactly the same as the proportions found in our solar system. The sort of things that were needed to make a planet or, say, a bat-eared fox.

Just think, the carbon atoms in the ink of the words you are now reading are older than the solar system. To quote from that old Woodstock song, we are stardust.

There are, of course, some very strange things out there. Star gazers have become very adept at working out how objects in space should move. A physicist named Fritz Zwicky was perplexed to find that the galaxies were moving so fast that clusters should be flying apart. Whatever it was that held them together had to have the gravitational pull of something far heavier than the galaxies themselves. That something was dubbed 'dark matter'.

Many pages of abstruse mathematics later, it was concluded that some 90 per cent of the universe was dark matter, and that the entities we call galaxies were no more than traces of sediment – mere archipelagos – trapped in vast swarms of invisible objects which held the whole show together.

Nobody yet knows what this stuff is, but at least some of the candidates are thought to be black holes.

The idea of a black hole has its roots in Einsteinian physics. What would happen if a huge neutron star collapsed with significant force to condense its core to a fraction of its size? The gravity would be so enormous that not even light could escape its pull. A black hole with the same mass as earth would, for example, be a mere marble nine millimetres across.

According to Einstein's calculations, not only would light be sucked into this monster but so would time – which would run backwards as it reached an 'event horizon'. If you could hold yourself at this horizon (this is sci-fi stuff!) the life of whole galaxies would pass before your eyes between one breath and the next. Huge black holes, great star-eaters, are thought to be at the heart of most galaxies

There are some cosmologists who speculate that black holes – the ghosts of dead stars – might create 'tears' in the space–time fabric, giving birth to whole new galaxies in unknown dimensions. In these terms our Big Bang may simply have been the birth moment of a black hole in another universe. The procreation of universes in different dimensional ecosystems? It sounds distinctly Darwinian.

Of one thing we can be sure, though: on a minor planet round an indifferent star in an unremarkable galaxy, a green fuse was lit and life began. The fine-tuning necessary for this to occur – over billions of years – is humbling.

The complexity of matter necessary to create a being able to reflect back on much of its inter-galactic origins, and forward to the time when its galaxy collides with another (as the Milky Way and Andromeda will do five billion years from now), is awesome.

Poets like Walt Whitman are permitted to say things like 'I know *nothing but* miracles'. Scientists have to be more cautious. Britain's Astronomer Royal, Sir Martin Rees, refers to the idea that the universe seems designed to produce life as 'anthropic reasoning'. But he says the chances of getting it right were about as likely as sitting at the bottom of a well and throwing a stone up so that it came to a halt *exactly* at the top.

Astronomer Dennis Sciama puts it rather more forcefully: 'Suppose you walk into a room and find a million numbered cards laid out on an enormous table. Suppose you find that these cards are placed progressively 1, 2, 3 and so on up to a million. Would you think they had been laid out at random, on the grounds that any particular ordering would have the same probability? Obviously you would not.'

Life on earth is the most complex and astounding thing we know. So it seems we can complete the journey to the beginning undaunted by Dave Laney's 'tyranny of the very large'. Colossal though they may be, stars and galaxies rank low on the scale of complexity. That is why it isn't over-presumptuous to aspire to understand them. A bat-eared fox poses a more daunting scientific challenge than a star. Statistically it's almost infinitely more extraordinary.

Dave had readings to take and my biorhythms hadn't adjusted to the all-night show. I crept down the dark stairs and into my car, making sure not to switch on the headlights. A kilometre down the dirt I spotted something in the road and slowed to a halt beside the bat-eared fox. A car had hit it and its beautiful, lifeless eyes were staring up into starry oblivion.

I rode on appalled. The whole universe seemed to be watching.

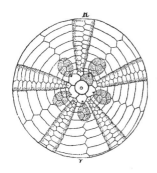

The strange history of a fix

If you'd like to drop in for a visit, my desk is S33°56'037" and E21°32'747".

Peering at these numbers on my hand-held GPS, I swear I can hear generations of sextant-clutching land surveyors weeping. The device has done to them what the motorcar did to wheelwrights and farriers.

It got me wondering, though, what I was 33° south of and 21° east of – and who figured it all out. Some digging led me to a story about a well in Syene (now drowned by the Nile's High Aswan Dam) and a brilliant Greek librarian and mathematician with the grand name of Eratosthenes. The year was around 200 BC, the date 21 June. On noon that day the sun shone straight down the deep well, leaving no shadows. Nothing so strange had ever been seen in the Egyptian city of Alexandria, where Eratosthenes lived.

In those days it was generally accepted that the earth was flat, and that the moon, sun and stars were fixed to the underside of a celestial sphere which rotated round the motionless earth. Who, indeed, could say they had felt the earth move?

The Greeks knew, however, that the sun inexplicably migrated north in summer and south in winter, and named the extremes of this migration 'tropics', after the word *tropos* meaning 'turn'. The midpoint between the turns they termed 'equator'.

Eratosthenes was fascinated by the story of the well, which a traveller had told him. It meant that at that time and place the sun was directly overhead, on what we know today as the Tropic of Cancer. But he had another idea: maybe the earth was round and somehow wobbled. After all, Socrates had noticed that ships sailing out to sea disappeared hull first and had suggested the possibility that the earth was a celestial ball.

If the sun's rays were parallel and the earth was a sphere, thought Eratosthenes, sunlight would hit different parts of the earth's surface at different angles. If he could discover the distance between Syene and Alexandria, then measure the angle of the shadow on the same midsummer day in Alexandria, he could use the triangle theorem of another Greek – Pythagoras – to measure the size of the earth.

Travellers gave him the rough distance between the well and his library in Alexandria, and the shadow gave him two sides of a triangle. He found the angle between the top of his pole and the tip of its shadow to be about one-fiftieth of a circle, so he simply multiplied the distance between Syene and Alexandria by 50 and declared the circumference of the earth to be 46,250 kilometres. He was only 6,000 kilometres out – and one must remember this was around 200 BC. Eratosthenes is rightly considered the father of modern cartography – the science of mapmaking.

For about 400 years this information lay in the great library of Alexandria, awaiting another genius who could comprehend its implications. That person was Claudius Ptolemy. His interests ran from biography and music to mathematics and optics, but his influence has come down through the centuries from two monumental books, one on astronomy and the other on geography.

He got many things wrong – it was AD 200 after all – but what he gave to the world was the subdivision of the degrees of a circle into minutes and seconds, and the notion that maps should be drawn with reference to these subdivisions as measurable coordinates. The task of the cartographer, he said, was to 'survey the whole in its just proportions'.

There is a parallel claim to this notion. The Chinese cartographer Chang Heng – a near contemporary of Ptolemy – wrote that a cartographer's job was to 'cast a network of coordinates about heaven and earth, and reckon on the basis of it'.

The source of his inspiration, it seems, came from a girl who was employed to embroider one of Chang's maps on silk. When he saw the intersecting lines of warp and weft, he was struck by their net-like effect on his map.

Eratosthenes had established three lines of longitude: the equator and the tropics of Cancer and Capricorn (named after star signs which coincided with summer and winter equinoxes). Ptolemy divided the circle of the planet as viewed from one of the poles into degrees, each of which was a line of longitude. It was he who originated the practice of placing north at the top, simply because most of the known world at the time was in the northern hemisphere and maps were easier to study with the land on top and all the writing and fanciful sketches below.

In the 500-year-old library of Alexandria, where both men had laboured, lay all the basic tools of modern cartography for the mapping of the planet. Then disaster struck. In the year 391 religious zealots sacked the library of Alexandria and burned its priceless contents, a symbolic victory of faith over reason.

We today cannot know what knowledge the world lost that day. Some works were hastily smuggled out, but most went up in flames, including much of the formulations of Eratosthenes and Ptolemy. Cartography was dealt a deadly blow. The earth, once again, became flat, the sun and stars were re-attached to their ever-circling heavenly sphere, and science died.

During the thousand-year slough of intellectual stagnation which followed – named the Dark Ages for good reason – surviving copies of the ancient cartographic texts were hidden by scholars and monks and handed down in secrecy. Maps reverted to being wildly speculative – more ecclesiastical than cartographic – and mostly dead wrong.

They sported the demons of the time: horse-footed men with ears so long they covered their bodies, savages who quaffed mead from human skulls, griffons with savage beaks, pythons which grew fat on the udders of cows, a cockatrice with crocodile forequarters and hindquarters suspended from lateral fins. It was thought that at the equator the seas would boil a boat to pieces.

The Renaissance and brave Portuguese mariners would change all that, dragging a slumbering Europe into a confrontation with the ideas of Greece's Golden Age and the exotic wonders of a New World.

Early in the fifteenth century Ferdinand Magellan circumnavigated the planet and it could never again be flat. The Portuguese drew maps; not good ones, but with usable information for adventurers to follow. And follow they did.

During the next century Eratosthenes's triangulation was dusted off and reinstated in the universities of Lisbon, and Ptolemy's notion of grids and degrees was widely debated in the salons and shipping offices of the great ports of Europe.

But the problem of how to fix a point on the earth remained

elusive. By what method could you know exactly how far you were from the equator, the pole or the nearest shore? Sailors died in the absence of this information. Explorers became hopelessly lost. The problem was that since the size of the earth was not known, the exact distance that a degree of latitude or longitude covered at any point was mere guesswork.

Enter Jean Picard. In 1669 this Frenchman – laboriously laying end on end well-seasoned, varnished wooden rods – measured an 11.4-kilometre baseline from Paris to Fontainebleau. Then, by Pythagorean triangulation, he sighted an arc from Paris to a clock tower in the village of Sourdon. Several other triangulations later he calculated – I will not tax you with his mathematics – that a degree of latitude measured 110.46 kilometres, a remarkably accurate result.

There was, however, a problem. Picard's measurement would be fine if the earth was a perfect sphere. But was it? Isaac Newton in England suggested that, because of centrifugal force, the earth bulged at the equator and flattened at the poles. The French Royal Academy of Sciences, for some reason, thought it was flatter round its waist and pointy at the poles.

To solve the problem, the Academy laid plans for two geodetic expeditions – one to Peru on the equator and the other to the Arctic Circle. They were to be terrible trips.

In 1736 French Academy member Pierre Louis Moreau de Maupertuis set off for Lapland with three other Academy members, a priest and a good deal of equipment.

The Swedish king tried to dissuade them from 'so desperate an undertaking'. But the party pressed on to the icy northern end of the Gulf of Bothnia where they made a base. Then they proceeded up the Torne River, braving cataracts and 'flies with green heads that fetch blood wherever they fix'.

Using methods pioneered by long-dead Eratosthenes, they triangulated their way across the Arctic Circle, laying baselines with chains and taking sightings on mountain peaks and known stars. To get accurate fixes they would climb peaks, strip the bark off trees to make them more visible, and build log pyramids. The effort amid the snow and ice was unthinkable. Fog and a forest fire they had started by accident delayed them.

After nearly two months of sightings they headed south, winter freezing the river behind them as they travelled. When it was all over, Maupertuis made his calculations. A degree, he found, was 111.094 kilometres, more than half a kilometre difference between a degree measured in France. The earth was definitely not a perfect sphere.

Lapland was almost a picnic compared to Peru. The other expedition, led by Pierre Bouguer and Charles Marie de La Condamine, encountered (as Bouguer wrote) 'difficulties not to be imagined'. The Peruvians and Spanish were suspicious: who could believe the ridiculous story about measuring arcs and fixing meridians? The party persevered amid hostilities, triangulating from Quito near the equator to a place named Cuenca in the Andes Mountains. Every angle measured was a testament to their endurance.

They hacked their way up the Andes, where members fainted from the altitude and had 'little haemorrhages' of the lungs. Clouds prevented sightings, rain fell for weeks on end. They lived on bad cheese and biscuits made from hard maize. One of the party, a surgeon, was killed in an argument in a village; a botanist suffered a mental breakdown and never recovered. The expedition took nine and a half years. La Condamine became the first foreigner to raft down the Amazon River.

His calculations showed that a degree of latitude at the equator was 109.92 kilometres, shorter than in Europe. The earth was definitely more curved at the equator and flatter at the Arctic Circle.

Given that the Arctic party had returned many years earlier, La Condamine arrived in Paris with incredible tales of the

Andes and Amazon, but no startling news. His weary lament: 'We returned seven years too late to inform Europe of anything new concerning the figure of the earth.'

In order to fix the imaginary planetary grid in place, only one line remained to be settled.

The equator is where it is because that position is half-way between the poles. It is the line of zero latitude, the prime parallel, and it could run nowhere else. But the line of zero longitude, the prime meridian, is another matter. It is arbitrarily designated and could be almost anywhere – and at one time or another it has been.

Ptolemy chose the Fortunate Islands, the westernmost extremity of his known world. Depending on the origin of the map, it has also run through Toledo, Cadiz, Madrid, Cracow, Copenhagen, Pisa, Naples, Rome, Augsburg, Ulm, Tübingen, Peking, St Petersburg, Washington, Philadelphia and Greenwich.

This caused immense confusion when pinpointing coordinates on maps, so in 1884 the United States government called an International Meridian Conference. At that meeting, with few dissenting votes, the Royal Observatory at Greenwich was chosen to be zero longitude. The grid was complete.

There is a sequel to this story which has to do with the GPS on my desk.

Satellites in space are now used as points for geodetic triangulation. Their use in global positioning systems makes cartographic calculations head-spinningly accurate. It has been confirmed that the earth is flattened, but not by much. Its equatorial diameter exceeds the polar diameter by a mere 42.77 kilometres.

But GPS accuracy has turned up something even more surprising, which would have got those old surveyors with their rods and chains really excited. It is also strangely charming.

Using instruments capable of measuring distances to within a couple of centimetres, plus 27 satellites and more than half a million optical, radio and laser sightings, the US Goddard Space Flight Center recently found that the southern hemisphere is larger than the northern hemisphere.

The earth is pear-shaped.

Th

God dam crazy

> *We're going to get their stinking dam. We've got secret plans. We're going to set up a laser beam below the dam, drill a tiny hole through the base of it. We've got underground chemists working on a formula for a new type of acid that will dissolve concrete underwater. We have suicide freaks who want to grow up to be human torpedoes. We've employed a crack team of serious Christians who are praying around the clock for an Act of God …*
>
> Edward Abbey, eco-activist

Dam engineers seem to think they've got a contract from God. Cornered the golden chair in the celestial faculty.

They've plugged the Tugela, the Orange, the Limpopo and the mighty Zambezi. They've dammed the Congo, the Niger and the Nile. The Mississippi has been halted, the Colorado no longer reaches the sea, the Ganges has been tamed and the Yangtze is a toothless dragon. They've pinched off the planet's arteries and pocketed the consultation fee. Every major river in Europe is not only stoppered like a bathtub but turned on and off like a tap. What's with these guys?

There are more than 40,000 large dams in the world and some 800,000 smaller ones. That's a shitload of dams, and every one of them is trashing the environment.

The Katse Dam, now bloating its belly in the Drakensberg, is so dangerous you're advised to change your holiday plans. And avoid China, give India a miss and for God's sake stay away from the Zambezi floodplain in the rainy season.

The mania to dam began after the Second World War and we're now running out of rivers, so we're re-damming dams. Just think about that.

Right now there is 10,000 cubic kilometres of water behind dam walls – five times the volume in the planet's rivers. Ten thousand frigging cubic kilometres! The weight is so great it triggers earthquakes, and geophysicists reckon the redistribution of weight from reservoirs may be having a measurable impact on the speed the earth rotates, the tilt of its axis and the shape of the gravitational field. Now that's outrageous!

But, leaving aside bobbing animals, drowned forests and pissed-off ants, there's something the dammers haven't been telling us: dams die. And when they do, all hell breaks loose. Many dams – especially hydroelectric mega-dams – are some of the most dangerous, ecologically invasive, morally corrupt and politically questionable structures human beings have ever built. It's time to start blowing them up. Seriously!

Let's start with dangerous. There are two main ways a dam can fail: 'overtopping' and foundation problems. Both can be caused by that dam-builder's nightmare: an earthquake.

One of the least-publicised effects of mega-dams is their tendency to *trigger* quakes. In their book *Vanishing Waters*, hydrobiologists Bryan Davies and Jenny Day explain that this is caused by water weight, bedrock lubrication and the pattern of filling. The mass of water in a dam like Kariba weighs, they calculate, around 180 billion tonnes. Cahora Bassa clocks in at 63 billion. That beggars belief.

This enormous mass, and the rate it accumulates during filling, force water into rock faults below the dam, lubricating them and causing them to groan and writhe. It's named, rather blandly, reservoir-induced seismicity.

As Kariba filled it caused two quakes higher than six on the Richter scale. Ever since, there have been rumours and leaked secret reports about structural damage to the dam's foundations. And it's not in a country with the wealth or political stability to do anything about it.

If Kariba goes, it will take Cahora Bassa with it. More than a million people will be in its floodpath. The towns of Tete and Quelimane would wind up in the Mozambique Channel. It'll make the recent floods – with babies being born in treetops – look like a picnic.

In a book named, poignantly, *Silenced Rivers*, Patrick McCully logged 32 dams hit by seismicity greater than four on the Richter scale.

If you want something to worry about, though, you need to go no further than the Katse Dam. This 187-metre monster (the tallest dam in Africa) is part of the six-dam Lesotho Highlands Water Scheme, one of the deepest and most sinuous dam systems in the world. When the project was signed into existence in 1987, no environmental impact assessment had been done. No checks, no balances. By 1997 the first phase of the scheme (the main aim of which is to get water to Gauteng) had run to US$9 billion. The Mohale Dam, the next stage of the project, will cost a further US$7.2 billion.

Within 18 months of the Katse filling there had been 95 seismic events in its vicinity. In February 1996 a tremor ripped a 1.7-kilometre crack in a mountain and through a village just north of the dam wall. Locals decided the Great River Snake was pissed off with the dam and got decidedly jumpy.

The Katse is engineered to withstand a 6.5 quake, but Chris Hartnady, formerly of the University of Cape Town, has suggested that a tremor as high as 7.1 could be on the cards because of the thinness of the earth's crust in that area.

Evidently there are no low-level sluices in this behemoth, so the water level can't be lowered if trouble looms. When the faeces hit the fan, the engineers in the hot seats will probably just run.

If the dam bursts when full, according to Davies and Day, the wall of water surging down the Malibamasto Valley will completely destroy Aliwal North several hundred kilometres away and still be five metres high by the time it reaches the Gariep Dam (the old H F Verwoerd Dam). Hang on, HF, your time is coming ...

When dams go, they really go big time. In August 1965 a storm hit Henan Province in China with a velocity estimated at a one-in-2,000-year event. The Banqiao Dam on a tributary of the Lower Yangtze filled close to maximum, but when its sluice gates were opened it was found that they were partly blocked by sediment. The dam kept rising. On 7 August it burst, and 500 million cubic metres of liquid hell exploded through the downstream valleys and plains at nearly 50 kilometres an hour. Entire towns disappeared. When the wall of water hit the Shimantan Dam it was flattened, as were 60 other smaller dams. The floodwaters formed a lake covering a thousand square kilometres. It is estimated that 230,000 people died, but the disaster was airbrushed out of history. Information about it leaked out only 20 years later.

In 1960 the Vaiont Dam – the world's fourth highest – was completed at the base of Mount Toc in the Italian Alps and began filling. Soon afterwards seismic shocks were recorded and a mass of unstable gunk began to slide down the mountainside towards the reservoir. The dam was partly drained and the shocks stopped. When filling recommenced the shocks returned, but engineers and geologists decided the slipping mass would keep moving slowly and filling continued. After heavy rains one night, 350 million cubic metres of rock broke off Mount Toc and plunged into the dam. A gargantuan wave – the height of a 28-storey building – overtopped the wall and within two minutes the town of Longarone, a kilometre downstream, was history. About 2,600 people died.

Dams are no more unbreakable than the *Titanic* was unsinkable.

If geographers can screw up on dam sitings and engineers can thumb-suck dam strengths, builders are no less fallible. Huge amounts of money are involved in dam construction and rackets abound. In 2001, 19 large, international contractors to the Katse project were in court on bribery charges. According to the charge sheets – and possibly with the knowledge of the World Bank (which provides most of the funds to build dams worldwide) – nearly US$2 million were handed to a Lesotho official in order to shoehorn the dam contracts into place. (According to the International Rivers Network, investigators have been denied access to certain World Bank files.)

In such an environment of graft, the temptation to make profits by cutting corners, using inferior materials, using under-strength concrete or turning a blind eye to expensive safety features is compelling. Short cuts by contractors building a dam on the Sand River in Mpumalanga, for instance, led to its recent collapse.

Worse. There is a growing suspicion that the World Bank could be using the leverage of vast sums of money to finance Third World dams in order to conduit funds to First World construction companies. The dams, it seems, may be incidental.

Two-bit politicians opening gigantic structures look good – not many will claim a dam isn't in the public interest. Developing nations end up with zillions in foreign debts and their leaders with dollar-lined pockets and dams named after them. And, as usual, dictatorial regimes are preferred in order to curb the inevitable backlash from dispossessed peasants. Who says colonialism's dead?

Dam failures, however, are only part of the dam problem. The Three Gorges Dam presently being built in China will displace more than two million people. The Narmada water project in India – being bitterly opposed by Arundhati Roy, the author of *The God of Small Things* – will boot out no less than 33 million. It is conservatively estimated that more than 65 million people worldwide, often from politically weak minorities, have been forcibly removed from their homes as a result of dam construction.

The damage dams do to the environment is incalculable. In *Silenced Rivers*, Patrick McCully describes each dam as 'a huge, long-term and largely irreversible environmental experiment without a control'.

Above the wall huge areas are simply inundated, often drowning so much greenery that intolerable amounts of methane and other greenhouse gases are pumped into the atmosphere from rotting biomass.

Davies and Day estimated that a single dam – Manyami in Zimbabwe – drowned several hundred tonnes of termites alone, never mind the earthworms, spiders, lizards, bacteria, fungi, thorn trees, grasses, daisies and mopane trees.

Very often more fertile lands are lost under dams than are gained through irrigation schemes from dams. When the Dneprostroi Dam was built in Russia, it inundated so much prime Ukranian farmland that Soviet hydrologists claimed burning hay harvested from the area it submerged could have yielded as much energy annually as was generated by the dam's power plant.

Downriver from dams, especially in Africa with its extensive flood and drought cycles, the damage to ecosystems is disastrous.

Floodplains – where rivers flow into the sea – are some of the world's most diverse ecosystems. Creatures great and small have evolved in synchrony with the flood cycles, nourished by the annual silt flows and dependent on the scouring effect of floods. Now, instead of flushing silt, the river simply deposits it in the dam. In high-silt regions after a number of years, the water is more akin to thick lentil soup. Crappy stuff on which to run a turbine. Downstream, everything dependent on nutrient-rich sediment starves.

Hailed as the solution to annual flooding, dams are simply killing floodplains – and people in them. Flood 'pulses' are the main reason for the astonishing diversity and productivity of floodplains, which have been calculated to be 65 times richer in life than the seas. The tropical seas near floodplains produce fish yields a hundred times higher than floodplain-free coastlines. In recent times, because of dams, around 20 per cent of the world's freshwater fish have become extinct, threatened or endangered.

Not only fish are dying, however. For years Bryan Davies warned of a disaster in the making in the Zambezi floodplain.

By taming the floods, the Cahora Bassa Dam forced peasant farmers to move into the floodplain and further upriver towards the dam. But they were not tamed that easily.

In 1998 Cahora Bassa tried to lower the dam by opening two sluices but harmonic vibrations forced their quick closure. Then in 2000 Cahora Bassa failed to respond to Kariba's releases and the area between the dams was flooded. In 2001 Cahora Bassa was forced to open four of its eight sluices, inducing a flood which killed 105 people and left 250,000 homeless. Today most of those who survived are still huddled in vast refugee camps south of Caia, on the brink of starvation.

Unbelievably, Zimbabwe now has plans to build another dam on the Zambezi – which will inundate the Batoka Gorge below Victoria Falls. Mozambique is planning to throw a wall across the Zambezi below Cahora at Meponda Uncua (and inundate the finest tiger fishing area in the world) so it can have its own mega-dam (curiously, Portugal owns Cahora Bassa). Don't these guys ever learn?

Are dams necessary? Of course: they provide water and power. But we don't need them so huge, in such large numbers and, often, in the countries in which they're built. They offer neither cheap nor clean technology.

Around eight litres in every ten of dam water go to agriculture. Most agriculture uses overhead watering systems which, especially in hot climates, loses most of the spray to evaporation. Only the accumulating salts stay in the soil. It's an expensive way to make clouds and a lousy way to grow wheat or maize.

Cities simply waste water: it's been estimated that 25 per cent leaks from crappy pipes before it gets to your tap. Even minimal water education backed by tough anti-waste laws would reduce urban industrial and home-use consumption by half. The battle cry shouldn't be more power, but better power and water conservation.

There can be only one conclusion: to halt dam construction and to blow up useless and dangerous dams. It is being done in the United States, and France has banned all further dam construction.

A good practice run in what has been termed 'dam decommissioning' could be the Gariep Dam or the spooky Vanderkloof Dam lower down the Orange (it's virtually an unmanned machine).

There are a number of ways you could do it. First you draw down the water as far as possible, then you begin lowering the wall with pneumatic hammers and explosives. Finally, when the danger of water surge is sufficiently reduced, you dynamite the sucker.

If downstream users are the sort who bitch about muck in their water – and they're bound to – you have to divert the flow, build coffer dams and blast sections of the wall inside them.

The big expense, especially if it's a mega hydro-dam with a 180-metre wall, is carting away all the rubble.

According to Bob Daphne of Wreckers Demolition, the cost will run to millions. But it could be done. 'Just show me the wall,' he responded.

Then again, a large amount of dynamite in the right place on some dark night could be spectacular. Strike a blow for Mother Nature. Blow a dam! Just don't kill anybody.

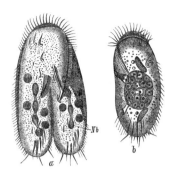

Just sniffing around

The mopane leaves smelt of ice cream. A wild foxglove gave off a nostalgic whiff of Cadbury's nut chocolate. One lowly pelargonium I tripped over emitted strawberry, another peppermint and a knobwood tree hinted of freshly squeezed orange.

A bit further down the trail some wild gardenia was doing things with perfume that was probably making a million butterflies swoon.

It hadn't occurred to me how aromatic the Bushveld could be. But the rains had so thoroughly clogged the scenery of northern Kruger Park with green stuff, I'd have stood on a lion before seeing it. So, during a hike, I started sniffing leaves for something to do; then got hooked on it and ended up having a nasal epiphany.

What was frustrating was my inability to put names to the exotic procession of aromas, and I soon realised it was because

English just doesn't have a vocabulary to adequately describe smell. In fact no language in the world does. We simply cannot utter the names of most of the thousands of chemicals faithfully picked up by the blunt instrument in the middle of our faces.

Even though an untrained nose can recognise thousands of smells, the few we can describe are based on our sense of taste: sweet, pungent, bitter. Or else we designate them by the things that emanate from them: coffee, paint, grass ...

In English the word 'smell' is forced into service for an odour, the process of perceiving that odour *and* the embarrassing business of emitting such a pong. It all seems very poor treatment of a sense without which food would taste like cardboard and a lover would have as much ethereal allure as a billy goat.

This vagueness was obviously intolerable to Carolus Linnaeus, that great warehouse clerk of nature who categorised and named every living thing he could get his hands on (because of him we're all described as *Homo sapiens*).

In his fever to catalogue all of existence, the obsessive Swede produced a study, *Odores medicamentorum*, in 1752, which today is so obscure no biologist I spoke to had ever heard of it.

There were, he decided after much sniffing around, seven major classes of odour: fragrant, aromatic, ambrosial, garlicky, goaty, foul and nauseating. Since then his classification has been tinkered with by chemists and perfumers but hasn't been bettered. Unfortunately, in our sanitised times, Linnaeus's work on aromas seems to have been simply buried.

Although the syntax of smell didn't catch on, aromas are an ever-present fact of life. Until quite recently, smell had far greater ramifications than today's billion-dollar perfume industry and the choice of toilet cleaners.

Up to the end of the nineteenth century cities of the West were stinking, poorly drained cesspools where the stench of tanneries and the sweat of the great unwashed mixed with the heady incense issuing from churches and the weedy tang of horse dung. Those who could afford to doused themselves with whatever perfumes they could get hold of. Smell was a persistent in-your-face issue and every social class held their noses in the presence of the class below it. The earlier Roman cities, if anything, were worse and Cairo was described by Napoleon as a stinking hellhole.

All this made people crave fanciful aromas. And dream they certainly did. Who, today, could describe a simple kiss, in the manner of the Roman writer Martial, as 'breath of balm from phials of yesterday, of the last effluence that falls from a curving jet of saffron, perfume of apples ripening, fields lavish with the leafage of spring'?

A thoughtful book, *Aroma*, written in the 1980s by three aromaphiles named Classen, Homes and Synnot, has an intriguing theory on smell.

'In the pre-modern West, odours were thought of as intrinsic "essences", revealing inner truth,' they write. 'Through smell, therefore, one interacted with interiors, rather than with surfaces, as one did with sight. Odours cannot be readily contained, they escape and cross boundaries. Such a sensory model is opposed to our modern, linear world view, with its emphasis on privacy, discrete divisions and superficial interactions.'

In a remarkable new book on smell – *Jacobson's Organ* – Lyall Watson is rather more blunt about the matter. Men, he says, can detect a woman ovulating from several blocks away. In the distant past, when our troglodyte ancestors coupled indiscriminately, that was all very well. But when humans began pair bonding, horny males sniffing the air were a danger to social harmony. These smells, says Watson, can seriously disrupt whole societies, so perfume was invented to disguise, rather than to enhance, sexual attraction.

Over the centuries, disuse and misuse have evidently led to the downgrading of our nasal sensitivity. Labradors, for example, have around 280 million olfactory sense cells, beagles more than

300 million and bloodhounds, those slobbering smell-seeker missiles, have an area of sensors about the size of this page. Yours, by comparison, cover about the size of a postage stamp.

Still, like it or not, our noses remain pretty sensitive and the air is loaded with suggestion. Even the cleanest air – say over Antarctica – has around 200,000 bits of this and that in every lungful. In the thick of a Serengeti migration or in rush-hour traffic, that could rise to two million.

Mostly it's minute particles of salt, clay and ash from forest fires or distant volcanic eruptions interspersed with exotic aeroplankton: a few stray viruses in transit between hosts, several stray bacteria, maybe 50 or 60 fungi including rusts and moulds, a number of algae, maybe a fern, moss or mushroom spore. Perhaps some airborne pheromones giving rise to wanton desires and the bittersweet chemical chatter of plants.

It's all on the odournet.

But to get back to the Bushveld trees in Kruger Park. Entranced by their delicious aromas, I confess I nibbled them but spat fast – they all tasted awful. So why did they have such heavenly aromas?

Plant geneticist Katherine Denby at the University of Cape Town's Microbiology Department seemed the right person to ask. She's in the arcane science of plant communication.

'They may smell good to you,' she commented, 'but for insects and other creatures these smells can be unpleasant and a warning. That's only the beginning of what's interesting about plant defence, though.'

The mustardy smell of vegetables such as broccoli, cabbage or radishes, it seems, is not to entice you but a chemical reaction to your attack on them. A scientific description of what occurs is weighed down by abstruse chemical formulae but, in short, what you smell is a herbivore toxin designed to scare you off.

Many plants resort to even tougher action. When attacked by

butterfly larvae, poplar trees give off aromatic phenols which inhibit the parasite's growth. Some plants, such as tomatoes, react to a bite by releasing toxins which prevent protein intake in the attacker's stomach, others inhibit the absorption of starch. Many flush their entire systems with bitter tannins which curb digestion and give the offending creature a bad stomachache.

'What I'm particularly interested in are plants which jasmonate,' Katherine informed me. When I looked blank she explained: 'When they're injured, many plants synthesise jasmonic acid, which is released into the air and warns other plants to take evasive action. They'll tanninise or do whatever's required to deter attack.'

It's been found that tannin levels can triple before the browsers even get to them. That's why leaf-munching animals keep moving – the trees simply won't allow them to chew the plant to death.

If there are too many browsers the whole area goes into aromatic red alert. It's even probable that different species of plant can pick up the jasmon danger signal and close ranks against the invaders. So who said plants can't feel?

Tobacco plants go even further. When attacked by mosaic virus they give off oil of wintergreen. Unaffected plants pick it up and convert the wintergreen to salicyclic acid, which is the main ingredient in aspirin. It's the nearest thing to inoculation.

Tobacco, together with cotton and maize, has another trick up its stalks. When caterpillars begin feeding on them, the plants send out an aromatic signal for help. This alerts a specialist parasitic wasp, which arrives to pluck the plant clean of crawlies.

'Plants are smarter than we think,' commented Katherine. 'All those smells you enjoy are their communication system. And we're only beginning to understand just how complex their relationships are.'

All this, I reflected as I trotted back down the stairs into University Avenue, is very bad news for vegetarians and people like me who keep tugging off leaves to smell them.

On the avenue I greeted a hedge, then untangled a branch which had been jammed down by someone's bicycle. When I patted the leaves affectionately, some students looked at me strangely. Too bad: I had bigger things to deal with. If that hedge blamed me for any bicycle damage to its leaves, the whole of Newlands Forest would get to know about it. I found that possibility rather intimidating. I wouldn't want the privet to send out the sort of signals that get me strangled by the bougainvillaea in the car park.

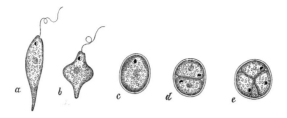

An enthusiasm for fire

'It's called myrmecochory,' said Sean Privett as I glanced at a battalion of ants tugging and rolling some seeds towards their hole. 'Seed dispersal by ants. These guys are doing fynbos fire duty.'

We'd just emerged from the cool interior of a milkwood forest he'd estimated at more than 1,000 years old. Its leathery leaves canopied to ground level in a perfect fire screen. Beyond the forest, unbroken fynbos of Grootbos Nature Reserve rolled down to the dune fields of Walker Bay Plain.

The Klein River Mountains, recently charcoaled by a huge blaze, dominated the horizon, and where their toes touched the ocean lay the crumpled-paper-looking clutter of Hermanus.

'This pincushion protea has sweet tips on its leaves to attract ants and keep them hanging round,' Sean continued, undeterred by my gawping at the view. 'When its seeds drop, the ants take them to their underground larders where they eat the tips and leave the seeds to wait for fire.'

The seeds in question were being worked over by pugnacious ants. They're well named and don't brook interference from a human finger. Sean, who's a plant ecologist, prodded a seed, then leapt back with a yelp, prising a pair of fiery pincers from the soft skin beside his nail.

'They eat anything,' he grumbled, then began hopping around as ants swarmed onto his boots.

'If you think having ant gardeners is a sophisticated arrangement,' he chuckled as we backed away from the increasingly aggressive seed grabbers, 'how about *Rovidula gorgonias*? It's a bit like a large sundew with sticky fingers which catch insects. But it does this only to feed a little blood-sucking bug which eats what the plant catches. The plant's reason for this good-neighbourliness is that it needs nitrogen and gets it from the bug's pee.'

I'd come to Grootbos with the rather vague wish to find out a bit more about fynbos. It's a biome with a reputation for diversity – books cite 8,600 species – but from a distance, and between flower seasons, fynbos looks pretty dull.

It didn't take much of Sean's tuition, however, to realise that you either have to take the long view and deal in millions of years, or get on your hands and knees with a magnifying glass. Either way you end up dealing with the biome's most significant element: fire. Without it, fynbos would wither and vanish.

The reason, it seems, is rooted in some wild and ancient geology and death-defying acts of self-immolation. I stopped staring at the horizon and began to pay more attention.

Cape fynbos has the highest species density in the world. It makes up four-fifths of the Cape Floral Kingdom, which stretches from Namaqualand to Port Elizabeth and encompasses the mountain ranges which run roughly parallel to the western and southern Cape coasts.

Around 5,800 of its plant species are endemic – an extraordinary situation generally found only on oceanic islands. By comparison the British Isles, three and a half times larger, has only 20 endemic plant species. Table Mountain alone – a mere 57 square kilometres – has 1,500.

This extravagance of life is completely incongruous – a wonderland in an area plagued by summer droughts, with soils leached of essential nutrients, where recurring fires blast everything in their path, gales howl for days on end and the amount of plant material produced is only marginally greater than in some deserts.

Fynbos diversity, we must assume, is not the product of a benevolent environment, but of adversity and stress. As anyone who has grabbed a piece of fynbos to steady themselves will know, it's tough, unforgiving stuff – and has good reason to be so.

The root of the problem, if we are to believe a remarkable book entitled *Fynbos* by Richard Cowling and Dave Richardson, lies in a tumultuous geological history. Around 1,000 million years ago – when the southern continents were all clumped together as Gondwanaland – the foundation of the present fynbos region was laid down as the shales, sandstones and limestones of the so-called Malmesbury Group.

After a relatively peaceful 500 million years, a spurt of mountain building and erosion began which produced a massive sandwich of sandstone and shale. This included the Table Mountain Group and a single layer of sandstone 3,000 metres thick – the biggest sand pile in the world.

Then, beginning 280 million years ago, massive chunks of sedimented sand, silt and clay – deposited in an earlier epoch – were thrust up into massive, crumpled pleats to form the Cape Folded Mountains. Time and weather then gradually eroded these, producing nutrient-poor sandy soils. Some 60 million years ago the great plant era began, with most of the world's landmasses carpeted with great forests.

In geological terms, the arrival of an ice age 35 million years ago was sudden. Temperatures dropped and so did sea levels. Wind-blown beach sand formed giant dune fields round the Cape coast as well as nutrient-leached sandy plains. This was tough on forests, but not on small proto-fynbos plants, which had skipped down the highland stepping-stones from their ancestral mountain homes in East Africa. And as they moved they got progressively stronger, developing a taste for innovation along the way.

Some 12 million years ago comes the first evidence of fires, and out of the ashes of the primeval rainforests the fynbos we know was born.

Lightning, sparks from falling rocks or perhaps even volcanoes would have started these blazes, but in the tinder-dry conditions at the end of dry summers they were inevitable.

Looking back from our vantage point in time it is difficult to know whether the fires engendered fynbos or whether fynbos caused the fires. All we can say is that, in some strange way, they now need each other. And having an adversary as powerful as fire has produced some of the smartest solutions in the plant kingdom.

Milkwood trees have evolved in response to fire, producing thick green fire skirts to ward off flames. One would think the many fynbos plants would take this approach. But most have gone the other way: virtually invoking fire and immolation.

A good number of fynbos plants pack their leaves with flammable resins and don't drop their branches, keeping them as fire-ready kindling. Their reproduction works on a 'burn thy neighbour' principle. When the fires come they immolate themselves and any plant around them in a fierce blaze, giving their fire-resistant seeds a better chance when the first rains fall.

Long before the arrival of humans, great walls of flame would sweep across the region with almost unfailing regularity. The last of these truly great fires occurred in February 1867. Having started in an unremitting berg wind near Swellendam, it finally died out a week later in the thickets near Uitenhage, some 500 kilometres to the east. By comparison, the recent Cape Peninsula fires were a mere sideshow.

After about 15 years most fynbos could be said to be just dying to burn. In the smouldering aftermath of a late summer fire, protea cones would have opened, releasing their precious crop of protein-rich seeds (which are pounced on by small rodents and buried by both them and ants).

First to appear above the ashes will be plants which have lain dormant as bulbs and others known as fire ephemerals. The spiky red balls of *Brunsvigias* literally shoot out of the blackened soil, followed (depending on where you are) by breathtaking displays of watsonias and fire lilies. These are spurred into action by smoke, which also conjures from the ground ericas, restios and other dormant creatures.

The first winter shower will filter precious, fire-released nutrients into the impoverished soil, prodding into action 100-year-old rootstocks, patient bulbs and stored seeds. Where there might have been ten species in an area before a burn, soon there will be 70 to 100. Post-fire fynbos literally brims with life.

There is an ancient Persian legend about a mythical bird named Simurgh. When it sees death draw near, it makes a nest out of sweet-smelling wood and resins. This it exposes to the full force of the sun's rays, which ignite it and the bird is burned to ashes. From its scorched marrow bones another soon arises to begin life anew.

The story's so apt it makes you wonder whether Persian travellers got as far as the Cape.

Leaving the ants to their seeds, we angled upwards for a bit until the fynbos around suddenly changed type.

'Now we're over limestone,' commented Sean. 'These plants like it.'

Pretty soon he was waving a rustly bit of bush under my nose. 'This erica is in the Red Data Book, it's very rare. It grows only here and on a hill near Hermanus.'

That's both a problem and a fascinating feature of fynbos: it's often micro-habitat specific. The reason for this – and for the high number of fynbos species – is again fire. Take the little pincushion named *Leucospermum prostratum*. It creeps along the ground and its blooms smell yeasty – just the way mice like them. The little rodents line up to sup nectar, getting their faces covered in pollen in the process and doing the job of pollination.

But seed dispersal relies on ants – and mice eat seeds. So the protea has to orchestrate pollination at a time when there are plenty of mice, and seed drops at other times when ants are abundant and mice aren't. It does this with consummate efficiency.

But after a fire the mice might have vanished, or all the ants may be crisped. With few seeds in the ground and no creatures to ensure the dispersal of seeds or pollen, *Leucospermum prostratum* could undergo a population crash, forcing it through a genetic 'bottleneck'.

Perhaps the altered genetic characteristics of a few surviving seeds will be imprinted in all future generations, giving birth to a new endemic species. In a land of adversity the best line of defence, it seems, is diversity and proliferation.

The only trouble is that, with such specific requirements and niches, fynbos has one of the highest numbers of Red Data Book plant species in the world – 1,406 at last count. Nearly 30 species are classified as extinct and 750 are threatened by alien encroachment. Less than one per cent of renosterveld – which used to cover most of the south-western Cape's coastal plain – is preserved. Most of the rest is under wheat.

If fynbos burns easily, aliens seem to burn even easier. And trees such as gum, hakea, pine, Australian acacias such as rooikrans and Port Jackson willow create a slow, hot burn and not a 'cold' fast one preferred by fynbos. So the heat aliens generate usually takes out everything around them, even underground seeds (except their own).

Still, there was certainly enough fynbos around Grootbos. As we pushed our way through head-high protea bushes towards the 4x4 on the jeep track way below, Sean gave one bush a shove and commented: 'The understorey's getting choked. What we need in this sector is a good burn.'

A few hours earlier I would have considered him to be mad, but instead I patted a struggling little *Leucadendron* on its cones and muttered: 'Don't worry, you'll soon be fried.'

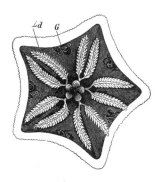

Of rain and reflections

Let me say this before rain becomes a utility we can regulate and sell: I celebrate its meaninglessness.

The rain I am in is not like the rain in the city. It fills the forest with a huge and confused sound. It hammers the corrugated-iron roof of the hut and stoep with insistent and controlled rhythms. I listen, because it reminds me again that the whole universe runs by rhythms I have not yet learned to recognise; rhythms not those of clock or engineer.

I came in from the city late this afternoon, sloshing up the track, and fired up some pasta and tomato sauce on the Cadac for supper. It boiled over while I was toasting some bread on the log fire. The night is now inky dark. The rain envelops the hut in its enormous, virginal myth – a whole world of meaning, of secrecy, of sudden silences, of rumour.

All that rainspeech pouring down, selling nothing, judging

nobody, drenching the thick mulch of dead leaves, soaking the trees, filling the streams and koppies of the woodland with water, washing out places where men have stripped the hillside.

What a thing it is to sit absolutely alone, in a forest, at night, cherished by this wonderful, unintelligible, perfectly innocent speech, the most comforting speech in the world, the talk that rain makes all by itself against the leaves, the talk of watercourses everywhere in the hollows, the shuddering expletives of thunder.

Nobody started it; nobody is going to stop it. It will talk as long as it wants, this rain. And I am going to listen as long as I am able.

Eventually I'm going to sleep, because here in this wilderness I've learned how to sleep again. Here I am not alien. The trees I know, the night I know, and the rain I know. I close my eyes and instantly sink into the whole rainy world of which I am a part, and the world goes on without me.

The city is filled with useful things. It is a *monument* to usefulness. Rows of houses, paved streets, electricity, shops, cellphone towers, trains; even helicopters that give you minute-by-minute traffic reports. There are schools for learning, hospitals for body and mind repair, and prisons for those who don't obey the rules. There are even clinics and beauty parlours for the animals we have tamed.

Unscheduled water is dealt with severely; guttered off roofs and streets, led into drains and captured in underground tunnels. Sometimes, when you cross a street, you can hear the water roaring below manholes down which the behaviour of the newly tamed river can be checked.

There is so little in the city that is not fabricated; if a tree gets among the apartments or even in a suburban street by mistake it is usually surrounded by paving stones or tar. It is given a precise location – a reason for existing – or it is yanked out. Or maybe it is left for tame dogs to pee on.

The celebration of rain cannot be stopped, of course, even in the city. The woman from the coffee bar scampers along the pavement with a newspaper over her head. A suited executive flips up his black umbrella and high-steps over a swollen gutter. The suddenly washed streets become transparent and alive, and the noise of traffic becomes a splashing of fountains.

You would think that urban citizens would have to take account of nature in its wetness and freshness, its baptism and its renewal. But though it waters gardens in the suburbs, the rain brings no renewal to nine-to-fivers, only tomorrow's weather and the glint of windows in tall buildings reflecting a turbulent sky. Somewhere inside those walls the 'real' city carries on, counting itself and selling itself with complex determination.

Meanwhile the citizens who *must* move plunge through the rain, intent on their busyness, slightly more vulnerable than before, probably piqued by the inconvenience of rain. Few see that the streets shine beautifully, that they themselves are walking on stars and water, that they are running over inverted skies to catch a bus or taxi, or to shelter in packed malls which ooze the mindless sound of elevator music.

But they must know there is wetness abroad. Perhaps they even *feel* it. For them it's a nuisance, nothing more. Am I presuming too much?

Sitting here in this hut, alone and perfectly happy, with wet sounds, a camp bed and a Cadac, I wonder at the organised pleasures marketed as vacation. Among the many things the city sells is the happiness of escaping from itself. It urges you to 'have fun' by presenting you with irresistible images of yourself as you would like to be: having fun without the interference of guilt or worrying about the cost. Having fun with others who are having fun.

If you flip through the back pages of a magazine like *Getaway* you'll find endless variations on a theme of how to take the conveniences and luxuries of the city out of town and into beautiful places. Hotels with courtesy fridges full of mind-altering liquors, safaris from luxurious tented camps with gun-slung rangers for 24-hour protection, advertisements which coo: 'Just come and relax.'

If you think I'm pushing a point, let me quote one taken at random: 'Luxury tents with en-suite shower and toilet, situated on a private game park, offers game drives, walks, horse trails and boat cruises. Beach trips, swimming pool, excellent cuisine, boma dinners, personal service. Children welcome.'

Nice holiday – undoubtedly a home-from-home atmosphere, with assurances that the conveniences of the city are right on hand. In practice, though, expensive fun often harbours a doubt, which can blossom into another full-blown need, which then calls for a still more credible and more costly refinement of satisfaction, which again generally fails you. The end of the cycle can be despair. How often do you feel the need for a vacation to recover from the stress of the one you've just had?

A person in modern society is generally a person in a rush, a person with no time, a prisoner of necessity who would find it difficult to understand that a thing might perhaps be without usefulness.

Those who dare to be alone in wild places, who celebrate and seek out times of solitude (are they a dying breed?) can come to see that the 'emptiness' and 'uselessness' of the wilderness which the collective urban mind fears and views from afar are necessary conditions for the sort of renewal the word 'holiday' suggests. It is derived from 'holy day'.

It is quiet reflection – moments of tranquillity which realign us with this shimmering planet – that renews us; not body comforts, en-suite bathrooms, sundowners and game drives.

It seems to me that it's only in these times of solitary simplicity that we can remind ourselves of our true capacity for maturity, freedom and peace.

Philoxenos, a sixth-century Syrian who left us some texts well worth digging up, said: 'I will not make you such rich men as to have need of many things. But I will make you true rich men who have need of nothing, since it is not he who has many possessions that is rich, but he who has no needs.'

Obviously we will always have some needs. But it is in times when we reduce our needs to the most simple, when we shed the material and mental clutter of the city, that we can see who it is we really are – since the only needs we then have are real ones. A rucksack, a sleeping bag, a Cadac and a pair of good boots. Some simple food.

The rain has stopped. The afternoon sun slants through the trees, refracting off myriad useless drops of water. The valley resounds with the totally incomprehensible talk of streams and fountains. Guinea fowl chatter and cluck in the clearing and a robin is scratching round in the useless underbrush beside the stoep.

There's nothing I would rather hear, not because these are better noises than other noises, but because they are the voices of the present moment. They remind me that I am utterly, solitarily, blissfully here.